The New Brain Sciences

Perils and Prospects

The last twenty years have seen an explosion of research and developments in the neurosciences. Indeed, some have called this first decade of the twenty-first century 'the decade of the mind'. An all-encompassing term, neuroscience covers such fields as biology, psychology, neurology, psychiatry and philosophy and include anatomy, physiology, molecular biology, genetics and behaviour. It is now a major industry with billions of dollars of funding invested from both public and private sectors.

Huge progress has been made in our understanding of the brain and its functions. However, with progress comes controversy, responsibility and dilemma. *The New Brain Sciences: Perils and Prospects* examines the implications of recent discoveries in terms of our sense of individual responsibility and personhood. Aimed at neuroscientists, psychologists, philosophers, medical professionals, students and the general public alike, it is an attempt to kick-start a discussion of where neuroscience is going.

Sir Dai Rees is Knight Bachelor, a Fellow of the Royal Society, an Honorary Fellow of the Royal College of Physicians and a founding Fellow of the Royal Academy of Medicine. He was President of the European Science Foundation (1993–9) and Secretary and Chief Executive of the UK Medical Research Council (1987–96).

Professor Steven Rose has been Professor of Biology and Director of the Brain and Behaviour Research Group at the Open University since the inception of the university in 1969. His research focusses on the cellular and molecular mechanisms of learning and memory.

The New Brain Sciences

Perils and Prospects

EDITED BY

DAI REES AND STEVEN ROSE

PUBLISHED BY THE PRESS SYNDICATE OF THE UNIVERSITY OF CAMBRIDGE
The Pitt Building, Trumpington Street, Cambridge, United Kingdom

CAMBRIDGE UNIVERSITY PRESS
The Edinburgh Building, Cambridge, CB2 2RU, UK
40 West 20th Street, New York, NY 10011–4211, USA
477 Williamstown Road, Port Melbourne, VIC 3207, Australia
Ruiz de Alarcón 13, 28014 Madrid, Spain
Dock House, The Waterfront, Cape Town 8001, South Africa

http://www.cambridge.org

First published 2004

Printed in the United Kingdom at the University Press, Cambridge

Typefaces Trump Mediaeval 9.5/15 pt. and Times *System* LaTeX 2_ε [TB]

A catalogue record for this book is available from the British Library

Library of Congress Cataloguing in Publication data
The new brain sciences: perils and prospects / edited by Dai Rees and Steven Rose.
p. cm.
Includes bibliographical references and index.
ISBN 0 521 83009 5 – ISBN 0 521 53714 2 (paperback)
1. Neurosciences – Philosophy. 2. Neurosciences – Moral and ethical aspects.
I. Rees, David A. II. Rose, Steven P. R. (Steven Peter Russell), 1938–
RC341.N53 2004
612.8′01 – dc22 2004045660

ISBN 0 521 83009 5 hardback
ISBN 0 521 53714 2 paperback

Contents

Contributors

Professor Patrick Bateson, The Provost's Lodge, King's College, Cambridge CB2 1ST, UK.

Patrick Bateson is Professor of Ethology, and Provost of King's College, Cambridge. He was formerly Director of the Sub-Department of Animal Behaviour at Cambridge and later Head of the Department of Zoology. He is currently Biological Secretary and Vice-President of the Royal Society. His research is on the behavioural development of animals, and much of his scientific career has been concerned with bridging the gap between the studies of behaviour and those of underlying mechanisms, focussing on the process of imprinting in birds. He has also carried out research on behavioural development in mammals, particularly cats. His most recent book is *Design for a Life* (with Paul Martin, 1999).

Professor Angus Clarke, Department of Medical Genetics, University of Wales College of Medicine, Heath Park, Cardiff CF4 4XN, UK.

Angus Clarke is a medically qualified geneticist and currently Professor in Clinical Genetics at the University of Wales, Cardiff. He has particular interests in the social and ethical issues raised by advances in human genetics and in the genetic counselling process, while also teaching and working as a clinician. He has worked on the clinical and molecular genetic aspects of ectodermal dysplasia, Rett syndrome and neuromuscular disorders. With his colleague Peter Harper, he wrote the book *Genetics, Society and Clinical Practice* (Harper and Clarke, 1997). He directs the Cardiff MSc course in Genetic Counselling.

Professor Paul Cooper, School of Education, University of Leicester, 21 University Road, Leicester LE1 7RF, UK.

Paul Cooper is Director of the Centre for Innovation in Raising Educational Achievement, University of Leicester School of Education. He is a chartered psychologist and was a schoolteacher for ten years before taking a succession of academic posts at the universities of Birmingham, Oxford and Cambridge. He is editor of the journal *Emotional and Behavioural Difficulties,* and was co-recipient of the 2001 *Times Educational Supplement*/NASEN Academic Book Award.

Dr John Cornwell, Jesus College, Cambridge CB5 8BL, UK.

John Cornwell has directed the Science and Human Dimension Project at Jesus College, Cambridge, since 1990. He is author of *The Power to Harm* (1996), and editor of *Nature's Imagination* (Oxford University Press, 1995) and *Consciousness and Human Identity* (Oxford University Press, 1998).

Professor Merlin W. Donald, Department of Psychology, Queen's University, Kingston, Ontario K7L 3N6, Canada.

Merlin Donald is Professor and Head of the Department of Psychology, Queen's University, Kingston, Ontario, Canada and Visiting Professor at the University of Toronto. A cognitive neuroscientist with a background in philosophy, he is the author of *Origins of the Modern Mind: Three Stages in the Evolution of Culture and Cognition* (Harvard University Press, 1991), and *A Mind So Rare: The Evolution of Human Consciousness* (Norton, 2001).

Professor Yadin Dudai, Department of Neurobiology, The Weizmann Institute of Science, 76100 Rehovot, Israel.

Yadin Dudai is Professor of Neurobiology at the Weizmann Institute of Science, Rehovot, Israel. His main research interest is in the processes and mechanisms of the acquisition and consolidation of items in memory. He serves on a number of national and international committees on science, science policy and education and is engaged in popularisation of science. His most recent book is *Memory from A to Z: Keywords, Concepts and Beyond* (Dudai, 2002).

Dr David Healy, North Wales Department of Psychological Medicine, Hergest Unit, Ysbyty Gwynedd, Bangor LL57 2PW, UK.

David Healy is Reader in Psychological Medicine in University of Wales College of Medicine and Director of the North Wales Department of Psychological Medicine. His books include *The Antidepressant Era* (Healy, 1998) and *The Creation of Psychopharmacology* (Healy, 2002).

Dr Helen Hodges, Department of Psychology, Institute of Psychiatry, King's College, De Crespigny Park, London SE5 8AF, UK.

Helen Hodges is Professor Emeritus of Psychology, King's College London and previously Head of Functional Assessment at ReNeuron Ltd. She has worked for the past fourteen years in the field of neural transplantation, assessing effects of primary fetal and cultured cell grafts in animal models of neurodegenerative and ischaemic brain damage. She was a founder member of ReNeuron Ltd, a biotechnology company dedicated to the development of stem cells suitable for repair of brain damage, with twin aims of generating safe and stable cells, and assessing their functional effects.

Professor Regine Kollek, FG Medicine/Neurosciences, University of Hamburg, Falkenried 94, D-20251 Hamburg, Germany.

Regine Kollek has a Ph.D. in molecular biology, but became increasingly involved in assessing technological risk. She is now Professor at the Research Centre for Biotechnology, Society and the Environment at the University of Hamburg. Her research focusses on the ethical and social implications of modern biotechnologies and aims to identify preconditions for compatibility of these scientific and technical developments with human dignity and social benefits. Her latest books are on genetic testing in health insurance and on pre-implantation genetic diagnosis. She is vice-chairperson of the German National Ethics

Council and a member of the International Bioethics Committee of UNESCO.

Professor Peter Lipton, Department of History and Philosophy of Science, Free School Lane, Cambridge CB2 3RH, UK.

Peter Lipton is Professor and Head of the Department of the History and Philosophy of Science at the University of Cambridge and a Fellow of King's College. His main philosophical interests lie in the theory of knowledge and the philosophy of science, especially questions concerning explanation and inference. He is also interested in topics in biomedical ethics and is a member of the Nuffield Council on Bioethics, having chaired its Working Party on Pharmacogenetics. He is the author of *Inference to the Best Explanation* (Routledge, 1991; 2nd edn, 2004).

Professor Alexander McCall Smith, Faculty of Law, University of Edinburgh, Old College, Edinburgh EH8 9YL, UK.

Alexander McCall Smith is Professor of Medical Law at the University of Edinburgh. He is vice-chairman of the Human Genetics Commission of the UK and a member of the International Bioethics Committee of UNESCO. He is the author of a number of books on criminal law and medical law, including *Errors, Medicine and the Law* (with Alan Merry), published by Cambridge University Press and, with J. K. Mason, *Law and Medical Ethics*, now in its sixth edition.

Dr Mary Midgley, 1a Collingwood Terrace, Newcastle upon Tyne NE2 2JP, UK.

Mary Midgley is a professional philosopher whose special interests are in the relations of humans to the rest of nature (particularly in the status of animals) in the sources of morality, and in the relation between science and religion (particularly in cases where science becomes a religion). Until retirement she was a Senior Lecturer in Philosophy at the University of Newcastle upon Tyne, where she still lives. Her best-known books are *Beast and Man, Evolution as a Religion, Science as Salvation, Wickedness* and, most recently, *Science and Poetry*.

Helen Pilcher, formerly of ReNeuron Ltd, 10 Nugent Road, Survey Research Park, Guildford, GU2 7AF, UK.

Helen Pilcher joined ReNeuron Ltd when it was founded in 1997, as a leading scientist in the cell biology group, tasked to make immortalised human stem cell lines. She then followed up her interests in public scientific debate as manager of the Royal Society 'Science in Society' team, and is now working as a writer for *Nature*.

Dr Lorraine Radford, School of Sociology and Social Policy, University of Surrey Roehampton, 80 Roehampton Lane, London SW15 5SL, UK.

Lorraine Radford is Reader in Social Policy at the University of Surrey Roehampton where she teaches courses on criminology and on violence. Her research interests and all of her publications are in the area of violence against women and children, particularly the impact of domestic violence upon children.

Sir Dai Rees, Ford Cottage, 1 High Street, Denford, Kettering NN14 4EQ, UK.

Dai Rees was previously President, European Science Foundation (1993–9); Secretary and Chief Executive, UK Medical Research Council (1987–96); Director, UK National Institute for Medical Research (1982–87). Prior to this he held various research and research management positions in industry, and was a lecturer in chemistry at the University of Edinburgh. He is a founding fellow of the Royal Academy of Medicine, a Fellow of the Royal Society and a member of Academia Europea.

Dr Iris Reuter, Department of Neuroscience, Institute of Psychiatry, King's College, DeCrespigny Park, London SE5 8AF, UK.

Iris Reuter is a consultant neurologist who is pursuing the clinical development of stem cell therapy by investigating graft efficacy and mechanisms in animals and supervising the assessment of Huntington's disease patients with primary fetal grafts.

Professor Hilary Rose, Department of Sociology, City University, 4 Lloyd Square, London WC1X 9BA, UK.

Hilary Rose is a feminist sociologist of science. She is Professor Emerita of Social Policy, University of Bradford, and Visiting Professor of Sociology at the City University, London. From 1999 to 2002 she was joint Professor of Physic at Gresham College, London. Books include *Love Power and Knowledge: Towards a Feminist Transformation of the Sciences* (Rose, 1994), cited as one of the 101 most important feminist books of the century, and (jointly with Steven Rose) *Alas Poor Darwin: Arguments against Evolutionary Psychology* (Rose and Rose 2000). Her current research interest is in biobanks, where genomics and social policy meet.

Professor Steven Rose, Brain and Behaviour Research Group, Department of Biological Sciences, The Open University, Walton Hall, Milton Keynes MK7 6AA, UK.

Steven Rose is Professor of Biology and Director of the Brain and Behaviour Research Group at the Open University, a position he has held since the inception of the university in 1969. His laboratory research focusses on the cellular and molecular mechanisms of learning and memory, with particular reference to Alzheimer's disease; a new edition of his prizewinning book *The Making of Memory* was published in 2003 and he is currently working on a book on 'the future of the brain.' In 2002 he received the Biochemical Society medal for excellence in public communication of science (see Rose, 2003).

The Rt. Hon. Lord Justice Sedley, Royal Courts of Justice, London WC2A 2LL, UK.

Stephen Sedley practised at the Bar for twenty-eight years, principally in the fields of civil liberties and discrimination law, until 1992 when he was appointed a judge of the Queen's Bench Division of the High Court. He became a Lord Justice of Appeal in 1999. He is an Honorary Professor of Law at Warwick University and the University of Wales at Cardiff, and Judicial Visitor at University College, London. He chaired the Judicial Studies Board's working party on the Human Rights Act 1998.

He is President of the British Institute of Human Rights and chair of the British Council's advisory committee on governance.

Dr Guido de Wert, Institute for Bioethics, University of Maastricht, 3015 GD Rotterdam, The Netherlands.

Guido de Wert is Professor of Biomedical Ethics at the University of Maastricht, The Netherlands, and a member of the Standing Committee on medical ethics and health law of the Netherlands Health Council. His main research interest lies in the ethics of transplantation and reproductive medicine and genetics. His publications include *Ethics and Genetics: A Workbook for Practitioners and Students* (Berghahn, 2003).

Professor Barbro Westerholm, Swedish Association of Senior Citizens, Box 22574, S-104 22 Stockholm, Sweden.

Barbro Westerholm, M.D., has among other things been Adjunct Professor of Drug Epidemiology, Karolinska Institute, Director General of the Swedish National Board of Health and Welfare, Member of Parliament (Liberal Party) and Chairman of the Parliamentary Committtee on Research Ethics. At present she is President of the Swedish Association of Senior Citizens, vice-chairman of the EU organisation AGE and chairman of the Board of Växjö University.

Part I Introduction: the new brain sciences

Introduction: the new brain sciences

STEVEN ROSE

THE RISE OF NEUROSCIENCE

The US government designated the 1990s as 'The Decade of the Brain'. The huge expansion of the neurosciences which took place during that decade has led many to suggest that the first ten years of this new century should be claimed as 'The Decade of the Mind'. Capitalising on the scale and technological success of the Human Genome Project, understanding – even decoding – the complex interconnected web between the languages of brain and those of mind has come to be seen as science's final frontier. With its hundred billion nerve cells, with their hundred trillion interconnections, the human brain is the most complex phenomenon in the known universe – always of course excepting the interaction of some 6 billion of such brains and their owners within the socio-technological culture of our planetary ecosystem.

The global scale of the research effort now put into the neurosciences, primarily in the United States, but closely followed by Europe and Japan, has turned them from classical 'little sciences' into a major industry engaging large teams of researchers, involving billions of dollars from government – including its military wing – and the pharmaceutical industry. Such growth cannot be understood in isolation from the social and economic forces driving our science forward.

The consequence is that what were once disparate fields – anatomy, physiology, molecular biology, genetics and behaviour – are now all embraced within 'neurobiology'. But the ambitions of

The New Brain Sciences: Perils and Prospects, ed. D. Rees and S. Rose.
Published by Cambridge University Press.

these transformed sciences have reached still further, into the historically disputed terrain between biology, psychology and philosophy: hence the more all-embracing phrase: 'the neurosciences'. The plural is important. Although the 30 thousand or so researchers who convene each year at the vast American Society for Neuroscience meetings, held in rotation in the largest conference centres that the United States can offer, all study the same object, the brain, its functions and dysfunctions, they do so at many different levels and with many different paradigms, problematics and techniques. Inputs into the neurosciences come from genetics – the identification of genes associated both with normal mental functions, such as learning and memory, and the dysfunctions that go with conditions such as depression, schizophrenia and Alzheimer's disease. From physics and engineering come the extraordinary new windows into the brain offered by the imaging systems – positron emission tomography (PET), functional magnetic resonance imaging (fMRI), magnetoencephalography (MEG) and others – acronyms which conceal powerful machines offering insights into the dynamic electrical flux through which the living brain conducts its millisecond by millisecond business. From the information sciences come claims to be able to model computational brain processes – even to mimic them in the artificial world of the computer.

Small wonder then that, almost drunk on the extraordinary power of these new technologies, neuroscientists have begun to lay claim to that final terra incognita, the nature of consciousness itself. This of course is to suggest that there is some agreement about how such an explanation of consciousness should be framed. But there is not. The rapid expansion of the neurosciences has produced an almost unimaginable wealth of data, facts, experimental findings, at every level from the submolecular to that of the brain as a whole. The problem is of how to weld together this mass into a coherent brain theory. For the brain is full of paradoxes. It is simultaneously a fixed structure and a set of dynamic, partly coherent and partly independent processes. Properties – 'functions' – are simultaneously localised and delocalised, embedded in small clusters of cells or aspects of working of system as a whole.

Anatomists, imaging individual neurons at magnifications of half a million or more, and molecular biologists locating specific molecules within these cells see the brain as a complex wiring diagram in which experience is encoded in terms of altering specific pathways and interconnections. Electrophysiologists and brain imagers see what, at the beginning of the last century, in the early years of neurobiology, the pioneering neurophysiologist Charles Sherrington described as 'an enchanted loom' of dynamic, ever-changing electrical ripples. Neuroendocrinologists see brain functions as continuously being modified by currents of hormones, from steroids to adrenaline – the neuromodulators that flow gently past each individual neuron, tickling its receptors into paroxysms of activity. How can all these different perspectives be welded into one coherent whole, even before any attempt is made to relate the 'objectivity' of the neuroscience laboratory to the day-to-day lived experience of our subjective experience?

Is this even possible? Most neuroscientists are committed to, at the least, a psychophysical parallelism of brain and mind, and in its strongest form a fully fledged reductionist collapse which sees mind as merely the epiphenomenal product of brain. This leaves some little local difficulties, such as reconciling objective third-person data about brain states with the subjective experience that philosophers refer to as qualia, to say nothing of resolving age-old paradoxes of free will and determinism. And the hard fact remains that at the end of the Decade of the Brain, and already some way into the putative Decade of the Mind, we are still data-rich and theory-poor. For some neurotheorists, there is no real problem. Truth, ultimate explanations, lie embedded in the molecular constituents of the nervous system, and molecular biology and the new DNA technologies, will eventually be able to offer full explanations, which will collapse or dissolve the problems faced by physiologists, brains mappers and even psychologists. This is the reductionist agenda, whose full philosophical and technological flowering is celebrated in popular books and media accounts as well as in our own journals and conferences. Is such reductionist confidence justified? Or are there 'higher-level' explanations of brain and mind processes that are irreducible? This dilemma remains central to many of our debates.

THE PROMISE OF NEUROTECHNOLOGY

But our knowledges, fragmented as they are, are still formidable. Knowledge, of course, as Francis Bacon pointed out at the birth of Western science, is power. Just as with the new genetics, so the neurosciences are not merely about acquiring knowledge of brain and mind processes but about being able to act upon them – neuroscience and neurotechnology are indissolubly linked. This is why developments occurring within the neurosciences cannot be seen as isolated from the socio-economic context in which they are being developed, and in which searches for genetic or pharmacological fixes to individual problems dominate. Such searches both celebrate and reinforce the simplistic reductionist agendas of neuroscience and neurotechnology.

It is clear that the burden of human suffering associated with damage or malfunction of mind and brain is enormous. In the ageing populations of Western industrial societies, Alzheimer's disease, a seemingly irreversible loss of brain cells and mental function, is an increasing burden. Risk factors for the disease include possessing an inappropriate form of certain genes, and a variety of environmental hazards; treatment is at best palliative. Huntington's disease is much rarer, and a consequence of a single gene abnormality; Parkinson's disease is more common, and now the focus of efforts to alleviate it by various forms of genetic engineering.

But whilst such diseases and disorders are associated with relatively unambiguous neurological and neurochemical signs, there is a much more diffuse and troubling area of concern. Consider the worldwide epidemic of depression identified by the World Health Organisation (WHO) as the major health hazard of this century, in the moderation – though scarcely cure – of which vast tonnages of psychotropic drugs are manufactured and consumed each year. Prozac is the best known, but only one of a myriad of such drugs, designed to interact with one of the brain's key neurotransmitters, serotonin. Questions of why this dramatic rise in the diagnosis of depression is occurring are rarely asked perhaps for fear it should reveal a malaise not in the individual but in the social and psychic order. Instead, the emphasis is overwhelmingly on what is going on within a person's brain and body. Where drug treatments have hitherto been empirical, neurogeneticists are offering to identify specific genes that might precipitate the condition, and in combination with the pharmaceutical industry to design tailor-made ('rational') drugs to fit any specific individual through what is coming to be called psychopharmacogenetics.

But the claims of the neurotechnologies go far further. The reductionist fervour within which they are being created argues that a huge variety of social and personal ills are attributable to brain malfunctions, themselves a consequence of faulty genes. The authoritative US-based *Diagnostic and Statistical Manual* now includes as disease categories 'oppositional defiance disorder', 'disruptive behavior disorder' and 'compliance disorder'. Most notoriously, a disease called 'attention deficit/hyperactivity disorder' (AD/HD) is supposed to affect up to 10% of young children (mainly boys). The 'disorder' is characterised by poor school performance and inability to concentrate in class, or to be controlled by parents. The 'disorder' is supposed to be a consequence of disorderly brain function associated with a particular neurotransmitter, dopamine. The prescribed treatment is an amphetamine-like drug called Ritalin. The WHO has drawn attention to what they perceive as an increasing worldwide epidemic of Ritalin use. Untreated children are said to be likely to be more at risk of becoming criminals, and there is an increasing literature on 'the genetics of criminal and anti-social behaviour'. Is this an appropriate medical/psychiatric approach to an individual problem, or a cheap fix to avoid the necessity of questioning schools, parents and the broader social context of education?

The neurogenetic–industrial complex thus becomes ever more powerful. Undeterred by the way that molecular biologists, confronted with the outputs from the Human Genome Project, are beginning to row back from genetic determinist claims, psychometricians and behaviour geneticists, sometimes in combination and sometimes in competition with evolutionary psychologists, are claiming genetic roots of areas of human belief, intentions and actions long assumed to lie outside biological explanation. Not merely such long-runners as intelligence, addiction and aggression, but even political tendency, religiosity and likelihood of mid-life divorce are being removed from the province of social and/or personal psychological explanation into the province of biology. With such removal comes the offer to treat, to manipulate, to control. Back in the 1930s, Aldous Huxley's prescient *Brave New World* offered a universal panacea, a drug called Soma which removed all existential pain. Today's Brave New World will have a multitude of designer psychotropics, available either by consumer choice (so called 'smart drugs' to enhance cognition) or by state prescription (Ritalin for behaviour control).

These are the emerging neurotechnologies, crude at present but becoming steadily more refined. Their development and use within the social context of contemporary industrial society presents as powerful a set of medical, ethical, legal and social dilemmas as does that of the new genetics, and we need to begin to come to terms with them sooner rather than later. To take just a few practical examples: if smart drugs are developed ('brain steroids' as they have been called), what are the implications of people using them to pass competitive examinations? Should people genetically at risk from Alzheimer's disease be given lifetime 'neuroprotective' drugs? If diagnosing children with AD/HD also really did predict later criminal behaviour, should they be drugged with Ritalin or some related drug throughout their childhood?

NEUROETHICS AND HUMAN AGENCY

More fundamentally, what effect do the developing neurosciences and neurotechnologies have on our sense of individual responsibility, of personhood and of human agency? How far will they affect legal and ethical systems and administration of justice? How will the rapid growth of human brain/machine interfacing – a combination of neuroscience and informatics (cyborgery) change how we live and think? These are not esoteric or science-fiction questions; we aren't talking about some science-fiction prospects about human cloning, but prospects and problems that will become increasingly sharply present for us and our children within the next ten to twenty years.

The editors of this book believe that it is vital both to help clarify the thoughts of the neuroscience community itself concerning these questions, and also to make what we currently know and don't know about the brain and its workings accessible to a wide public in sufficient detail to kick-start a discussion of where our science is going, and above all of its medical, legal, ethical and social aspects. That these concerns are shared by many is indicated by the way in which yet another neologism, 'neuroethics', has emerged over the last couple of years, with professional ethicists and philosophers contributing to a vigorous discussion both within professional journals and at especially convened meetings.

The papers that form the chapters of the present book emerged as a result of two such meetings, held in 2001 and 2002. The first, 'Perils and Prospects of the New Brain Sciences', was convened jointly by the Wenner-Gren and European Science Foundation and took place at the Wenner-Gren Centre in Stockholm; the second, on 'Science and Human Agency', was a joint meeting of the Royal Society and Gresham College, in London. The two complementary meetings involved a range of presentations from many disciplinary perspectives, law, sociology, ethics, education, psychology, neuroscience, genetics and psychiatry. As editors, we have encouraged a subset of the speakers at these meetings to develop their presentations into fuller papers, and have then edited and reordered them so as, we hope, to make them as accessible as possible to as wide a non-specialist audience as possible, and we wish to pay tribute to the cooperation of our authors in submitting to this procedure. The contributors to the original meetings, and those whose chapters appear here, were chosen for their known critical expertise; you will find no gung-ho overoptimistic forecasts of the wondrous cornucopia of benefits that neuroscience might bring here. We are all too well aware of the overselling of the technological promise of the new genetics that began in the late 1970s. Nor, though, are our authors doom-sayers with an almost automatic rejectionism in response to new findings.

THE PLAN OF THE BOOK

The resulting sixteen chapters between this Introduction and the final summarising one, by Dai Rees and Barbro Westerholm, fall into three broad sections. The first, comprising five chapters, we have called 'freedom to change'. Here we focus on the extent to which current findings in neuroscience might cause us to revise our classical ideas about human consciousness, free will, determinism, agency and responsibility. The first chapter, by the philosopher Mary Midgley, sets the scene by asking how free we are to 'really' act? The psychologist Merlin Donald then considers the emergence of human mind and consciousness from within an evolutionary perspective.[1] He argues that the key features in the emergence of human consciousness lie in the nature of humans as social animals, but that mind and consciousness are not so much the property of individual brains but an expression of a relationship of the individual person with the social world in which that person is embedded. This theme is taken up by the feminist sociologist Hilary Rose, who looks with a degree of wry scepticism at the claims of neuroscience to appropriate consciousness from the other discourses – including those of the novel – in which it has featured over many years. Professor of technology assessment Regine Kollek, who has had a long-standing concern with developments in gene technology, revisits some of these concerns in the first section

[1] Merlin Donald was prevented from attending the Stockholm conference as his flight from Canada was blocked in the immediate aftermath of the attacks on the World Trade Center on 11 September 2001. This chapter is based on the talk he had intended to give.

of the book in the context of some of the claims of neuroethics. In particular, reinforcing Merlin Donald's and Hilary Rose's arguments, she contests the current neuroscientific attempt to reduce the concept of self to 'nothing but a bunch of neurons'. Lastly in this section, philosopher Peter Lipton turns once more to the classical questions: to what extent does neuroscience resolve traditional dilemmas of free will versus determinism, of human agency? Yes, Lipton insists, the determinism/free will dilemma is a false one emerging more from philosophical lack of clarity than from any advances in the brain sciences.

The second section turns to questions of human responsibility (agency) and the law. To what extent have the neurosciences affected our sense of responsibility for our actions, and in particular the traditional legal concept of *mens rea*? Might it be feasible to argue for instance, diminished responsibility for a criminal act on the grounds of genetic predisposition? Certainly this defence has been tried in the United States (Patrick Bateson refers to it in passing in his chapter as the 'Twinkie defence'). Professor of medical law Alexander McCall Smith, whose service on the Human Genetics Commission and the Nuffield Council's inquiry into the implications of behaviour genetics has given him a special concern with these questions, reviews the current principles involved in the concept of responsibility in law and how these might be affected by scientific advance. His paper is complemented by the practical perspective on how courts treat evidence for responsibility provided by one of Britain's leading judges, Lord Justice Stephen Sedley. Feminist sociologist Lorraine Radford analyses the evidence advanced both by behaviour geneticists and evolutionary psychologists for a genetic base for human aggression, and most specifically for violence by men directed at women, revisiting some of the issues raised in McCall Smith's chapter and their implications for governance. Lastly in this section the ethologist Patrick Bateson disentangles the tortured debate over nature and nurture, instinct and responsibility from a consideration of the processes of development.

These involve a continued dialogue between genes and environment, which precludes any simple assumption of genetic determinism.

The final section of the book turns to the stewardship of the new brain sciences in the context of more specific areas of current concern, where developments in neuropharmacology and genetics impinge most directly on social matters. We begin with a cautionary chapter by the neuroscientist Yadin Dudai, whose own research has been at the forefront of studies of the molecular mechanisms of memory formation. Dudai cautions against overoptimistic expectations for the immediate social and medical benefits that might flow from basic neuroscientific findings. Clinical geneticist Angus Clarke deconstructs the claims of behaviour genetics in one of its most prized areas, that of intelligence, or IQ. Turning to an even more controversial area, the next two chapters tackle the question of the actual and potential use of stem cells derived from human embryonic material. Such stem cell research and cloning (to be distinguished from cloning and bringing to term an actual human fetus, which is illegal) is now permitted in the UK and elsewhere in Europe. The potential of stem cells in the treatment of degenerative neurological conditions such as Parkinson's disease has been enthusiastically endorsed by some neurologists, but the ethical implications of their use have also raised widespread concerns. The cell biologist Helen Hodges has been at the forefront of research in this area, and writes with her colleagues Iris Reuter and Helen Pilcher to review the potential of such cells for a wide range of conditions. Her chapter is complemented by that of Guido de Wert, a professor of biomedical ethics serving on the standing committee on medical ethics and health law of the Netherlands Health Council, who considers the types of ethical and social objection that have been raised to the research and potential use of such cells before offering a (cautious) green light.

The next three chapters deal with drugs. Writer John Cornwell considers the widely prescribed specific serotonin reuptake inhibitor (SSRI) Prozac – in particular the legal attempt to hold its manufacturer,

Eli Lilly, responsible for the shooting spree embarked on by a man allegedly under the influence of a prescription dose of the drug. Clinical psychiatrist David Healy looks at the changing fashions in diagnosis of mental distress, from anxiety to depression to panic disorder, and their relationship to the ways in which the large drug companies promote their products, including the SSRIs. He calls attention in particular to the widespread practice of 'ghostwriting' apparently favourable trials of the drugs by leading and seemingly independent practitioners. The third chapter in this sequence turns to another widely prescribed drug, methylphenidate (Ritalin) whose widescale use is discussed earlier in this Introduction. Professor of education Paul Cooper considers the implications of the increasingly common diagnosis of AD/HD and the attendant prescription of Ritalin in British schools for educational practice.

The concern of all the contributors to the meetings that bred this book has been above all to raise awareness of some of the directions in which neuroscience is currently driving with a view to asking how both governments and civil society can respond proactively to potential developments, rather than retrospectively closing stable doors after the horses have bolted, as has happened so frequently in the history of the technosciences, from nuclear physics to genetics. Thus the final word is left to Dai Rees, former Chief Executive of the Medical Research Council (UK) and President of the European Science Foundation, and Barbro Westerholm, former chairman of the Swedish Parliamentary Committtee on research ethics and a pharmacologist by training, to suggest some lessons and implications.

ACKNOWLEDGEMENTS

In preparing this book for publication, the editors have been privileged to draw on the support of a number of people. The planning committee for the Wenner-Gren/European Science Foundation meeting included Konstantin Anokhin, Werner Callebaut, Regine Kollek and Carmen Sandi; that for the Gresham College/Royal Society meeting included

Patrick Bateson and Hilary Rose. The first meeting was made possible by generous funding and administrative support from both Wenner-Gren and the European Science Foundation and we would like in particular to thank Torvard Laurent and Marianne Minkowski. Funding for the second meeting was provided by Gresham College and support in kind from the Royal Society.

Part II **Freedom to change**

1 Do we ever really act?

MARY MIDGLEY

WHAT IS REDUCTIONISM?

Reductionism comes in two phases. First, there is the monistic move where we explain a great range of things as only aspects of a single basic stuff. Thus, Thales says that all the four elements are really just water. Again, Nietzsche says that all motives are really just forms of the will to power, and Hobbes says that mind and matter are both really just matter. Second, there can follow the atomistic move – made by Democritus and the seventeenth-century physicists – which is slightly different. Here we explain this basic stuff itself as really just an assemblage of ultimate particles, treating the wholes that are formed out of them as secondary and relatively unreal. (I have discussed the various forms of reductionism more fully in Midgley (1995).)

Both these drastic moves can be useful when they are made as the first stage towards a fuller analysis. But both, if made on their own, can be absurdly misleading. It is pretty obvious that Nietzsche's psychology was oversimple. And, if we want to see the shortcomings of atomism, we need only consider a botanist who is asked (perhaps by an archaeologist) to identify a leaf. This botanist does not simply mince up her leaf, put it in the centrifuge and list the resulting molecules. Still less, of course, does she list their constituent atoms, protons and electrons. Instead she looks first at its structure and considers the possible wider background, trying to find out what kind of tree it came from, in what woodland, in what ecosystem, growing on what soil, in what climate, attacked by what predators, and what had been happening to the leaf since it left its tree. And so on.

The New Brain Sciences: Perils and Prospects, ed. D. Rees and S. Rose.
Published by Cambridge University Press.

This holistic approach is not 'folklore'. It is obviously as central and necessary a part of science as the atomistic one. So it is odd that, at present, people seem to entertain a confused notion that science is essentially and merely reductionist, in a sense that includes both the kinds of simplification just mentioned, while 'holism' is an old-fashioned superstition. This reductionism is an intellectual fashion, that surely survives quite as much by its imaginative appeal as by its arguments. In a confusing world, such a picture of knowledge as modelled on a simple, despotic system of government, is highly attractive. I think it is no accident that the reductive method had its first great triumphs in the seventeenth century, at the time when the wars of religion filled Europe with terrifying confusion. The monolithic reductionist pattern seemed able to impose order on this chaos just as Louis XIV and the other despotic rulers of the time did. This was a style that accorded with the religious and political notions of the time.

In politics, that simple vision no longer commands much respect today. But in the intellectual world it has not yet been fully discredited. There, it seems to offer order and simplicity – which are, of course, entirely proper aims for science – at a low cost, avoiding the complications that often make it so hard to achieve these ideals. If we want to show that this cheapness is illusory, I think that we need striking images. I have already proposed one such image (Midgley, 1996, 2001), and I shall develop it a little more here.

DIFFERENT QUESTIONS, DIFFERENT MAPS

My image is that of the relation between different maps of the same territory in an atlas. Why (we may ask) do we need so many different maps of the world at the beginning of our atlases? Surely there is only one world? And why, in particular, do we need a political map of Europe as well as a physical one? Surely all real facts are physical facts? There are no separate, ghostly political entities. Would it not (then) be much more rational to represent everything political in physical terms?

After all, the countryside does not become suddenly different at the point where France meets Germany. Is it not superstitious to represent this frontier by a black line and a quite different colour? Any barriers that may mark it are ordinary physical objects of a kind that occur elsewhere, and, if necessary, they can be given their own symbols on the map. Similarly, towns are physical objects, consisting of items such as houses and streets and people. The frequency of all these items can, if necessary, be indicated by special colours. And – if we want more detail – their physical nature can be investigated through increasingly detailed research programmes. At this point, however, we cannot avoid asking, what is our principle of selection? The map is growing more and more confusing, yet it still does not tell us what we need, if we want to know (for instance) what language and what laws we are going to have to deal with here. Neither does it record everything physical. Why are we picking on certain objects for depiction rather than others? Why should the map concentrate on towns and roads and frontiers, rather than on climate or vegetation or the nature of the soil? Why mark populations of people and houses rather than populations of voles or dandelions or quartz crystals or slime moulds or bacteria?

The truth is, of course, that no map shows everything. Each map concentrates on answering a particular set of questions. Each map 'explains' the whole only in the sense of answering certain given questions about it – not others. Each set of questions arises out of its own particular background in life – out of its own specific set of problems, and needs answers relevant to those problems.

The arrangements that we see on political maps – frontiers, provinces, capital cities and so forth – show us the answers that are currently being given to certain social questions, questions that arise out of the difficulties of organising human life. When we look at these maps, what we want is to know the present state of these answers. There is no way in which the answers given to quite different questions about the physical constitution of the world could stand in for this knowledge, any more than finding shelter can stand in for finding food. Each set of questions comes in from a different angle and needs a different kind of solution. Of course the different sets can often be connected, which is why our view of the world is not utterly chaotic. But there is no way in which they can be piled up into a sequence where one set of answers translates or supersedes the others.

Physics is not 'fundamental' to the other enquiries in the sense of revealing a deeper reality, a final explanation for all the other kinds of problem. It is simply one kind of abstraction among others. Physics, like the North Pole, is a terminus. It is where you should end up if you are making one particular kind of enquiry – namely, a physical one. It is no help if you are trying to go somewhere else.

ABSTRACTION, ELEGANCE AND THE UNDERGROUND

John Ziman (1978) has given a pleasing extra twist to this imagery of maps by considering the map, or rather diagram, that is usually displayed for the London Underground Railway. This map depicts a beautifully neat system, reducing the painful chaos of the real city to a few straight lines and simple angles. Just as many people are inclined to think (at an imaginative level) that physics can reduce the muddled living world to an idealised mathematical form.

Does this map of the Underground perhaps present us with (forgive the pun) a deeper reality? A recent novel by Tom Sharpe tells of a civil servant who does indeed have a secret ambition to straighten out the city of London, making it conform to the idealised map of the Underground . . . This strikes me as an echo – though of course an absurd one – of the way in which philosopher David Chalmers (1995) has suggested that we should view the relation of physics to consciousness:

> Biological theory has a certain complexity and messiness about it, but theories in physics, insofar as they deal with fundamental principles, aspire to simplicity and elegance. The fundamental laws of nature are part of the basic furniture of the world, and physical theories are telling us that this basic furniture is remarkably simple. If a theory of consciousness also involves fundamental principles, then we should expect the same.

By proposing that the structure of consciousness must be fundamentally simple in the same way as physics, Chalmers tries to abstract consciousness from the complexity and messiness of life. But life is, in fact, the only context in which we know that consciousness can occur at all. To the contrary, it might be just as plausible to think of consciousness as simply an intensification of life – a stronger form of that power to use and respond to one's surroundings which is characteristic of all living things. This would be liable to suggest that consciousness – being yet more complex – is likely to be even messier than the rest of life.

In fact, the exaltation of reason is not itself always reasonable. The physics-envy that so often consumes biologists and social scientists today is no help to them.

The kind of difficulty which I am comparing to the problem of relating two maps can arise in many different contexts. It is found wherever two different conceptual schemes are used to describe the same phenomenon. In practical terms, it crops up whenever different agencies (such as the police and the probation service) have to tackle a single problem (for instance, juvenile crime or child abuse). And in theoretical terms it can arise between different branches of the various sciences wherever they share a topic. But there is one question on which at present all of us – not only specialists – are in constant trouble, and that is our understanding of the nature of our own actions and the actions of those around us.

Ought we to be explaining those actions in terms provided by the physical sciences, or are we still allowed to explain them in terms that are useful to the actors themselves? Can it be legitimate, in a world where physical science is as deeply respected as it is today, to go on construing human action in the non-scientific terms which allow us to understand them from the inside?

WHAT DOES ACTION MEAN?

The question is: do we ever really act? When we say that we have acted deliberately, rather than just drifting or being driven, we mean that we have done something on purpose. And when that happens, our actions can be explained – often with great success – by reference to our conscious thinking, especially our purposes. It is sometimes suggested, however, that this kind of explanation by purpose is unreal, that it is just an illusion. The true cause of our action is always a physical event, usually (of course) an event that we ourselves know nothing about. It follows that those apparently successful explanations by purpose are just mistakes.

We ourselves are (then) never really active agents at all. We are always passive, always *being driven – like people hypnotised or possessed by an alien force.* Indeed, this metaphor of driving was the one that Richard Dawkins used when he wrote that 'we are survival machines, robot vehicles blindly programmed to preserve the selfish molecules known as genes'.

Dawkins's language seems to imply, somewhat mysteriously, that the gene itself is a real agent, a kind of hypnotist doing the driving. It's not clear quite what he means by this. But the idea certainly is that the human being that we take ourselves to be is passive, not in charge, not effecting events by its thought. More persuasively than Dawkins, Colin Blakemore and other contemporary writers have suggested a similar arrangement in which agency is transferred, not to the genes but to the brain. Blakemore (1988) puts it like this (the added emphasis is mine):

> The human brain is a *machine which alone* accounts for all our actions, our most private thoughts, our beliefs. It creates the state of consciousness and the sense of self. It makes the mind . . . To choose a spouse, a job, a religious creed – or even to choose to rob a bank – is the peak of a causal chain that runs back to the origin of life and down to the nature of atoms and molecules . . . We feel ourselves, usually, to be in control of our actions, but that feeling is itself a product of our brain, whose machinery has been designed, on the basis of its functional utility, by means of natural selection . . . All our actions are products of the activity of our

> brains. It seems to me to make no sense (in scientific terms) to try to distinguish sharply between acts that result from conscious attention and those that result from our reflexes or are caused by disease or damage to the brain.

Here the brain seems to be personalised and credited as a distinct agent. The suggestion is that we should no longer say that we 'use our brains' or think with our brains, just as we say that we see with our eyes and walk with our legs. We should no longer consider the brain as one organ among others. Instead, we are now to consider ourselves as beings who are separate from it and are driven by it, as if it were a kind of hypnotist. This third-person agent is to displace the first person altogether from effective control of decision-making.

This is not just a trifling verbal change. It makes a difference to social life because, normally, the distinction between deliberate activity and mere passive drifting is often of the first importance to us. We need to know whether the people we deal with are in full charge of their actions or are in some way passive to outside forces – whether, for instance, they are drunk or psychotic or have just been blackmailed or hypnotised. We have to deal with them as conscious, responsible agents, not as mere whirring lumps of matter. Our notion of responsibility centres on a knowledge of people's purposes. And responsibility covers a much wider area of life than mere blame and punishment. It covers the whole ownership of actions, the notions that we form of people's characters, the grounds of our entire social attitude to them. In considering these things, we constantly pay attention to what we believe them to be thinking.

THE QUEST FOR SIMPLICITY

Now, since this centrality of the first-person point of view is a matter of common experience, theorists would obviously not have tried to eliminate such a crucial tool of thought if they hadn't thought they had a good reason for it. The reason that moves them is, pretty clearly, *a particular notion of what explanation is* – a conviction that

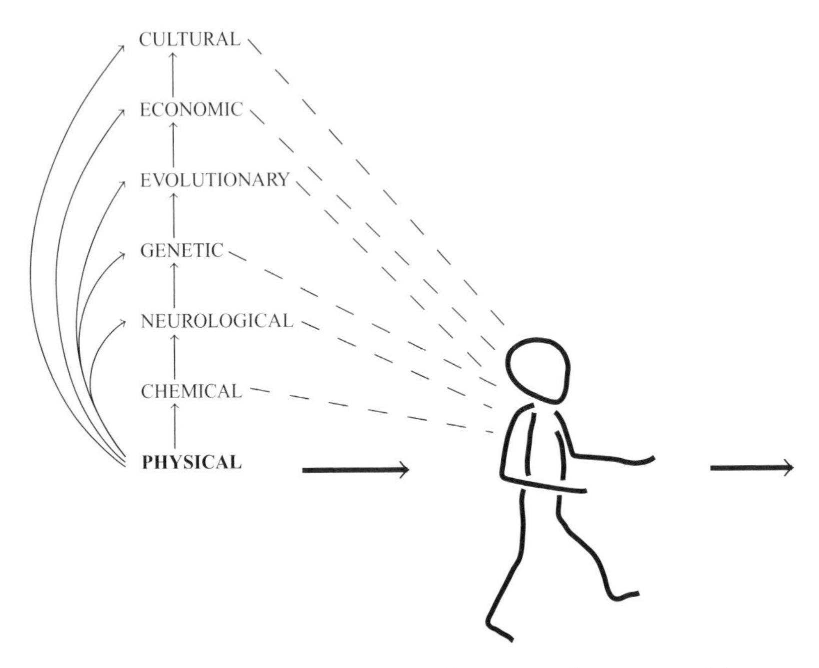

FIGURE 1.1. Causation: seventeenth-century reductionist model.

all explanation must be causal and that the only legitimate form of causal reasoning is one that cannot be extended to cover purpose.

That form is the simple linear model of causation that the founders of modern science, from Descartes to Newton, established in the seventeenth century (Figure 1.1). It still has a huge influence today because it is so beautifully simple. It attributes all real causation to collisions between the ultimate solid particles of which everything was then supposed to be made. Physics, which studied those particles, was therefore held to be the only study that could observe real causal interactions going on. All other kinds of apparent causation were just consequences of these atomic collisions and would eventually be explained in physical terms. Thus, all the other sciences, which traced those apparent connections, were only making provisional speculations at a superficial level. They dealt only in appearances and could always be mistaken. Physics alone could plunge down to the rock of reality. (This metaphor of *surface* and *depth, shallowness* and *solidity* is essential to the model's seduction. Nobody wants to fall through the floor, so the idea of a simple, infallible ultimate science was naturally welcome. And physics then seemed to offer that foundation.)

MANY QUESTIONS, MANY ANSWERS

Those were the days. Since that time, a whole raft of considerations have made it clear that we don't need that kind of metaphysical simplicity, and that we certainly can't have it. The most obvious reason for this is, of course, that we no longer have those handy ultimate solid particles, with their single simple habit of colliding. Clockwork machinery no longer rules today. Physics is now far more complicated. That is why physicists themselves tend now to be much less devoted to the old sweeping model than many biologists and social scientists are.

The deeper reason for the change is, however, that *we now have a much more realistic conception of what explanation itself involves*. We have begun to understand that the real world actually is complicated, and particularly that the people in it are so. Because they are complex, we need to ask many *kinds* of question about them, not just one. So we need to use many different ways of thinking in order to answer these questions.

That is the reason why we need to use many different disciplines. Their different ways of thinking are required for separate jobs. They are tools adapted to resolve distinct problems, not rival monarchs competing for a single throne. In fact, this isn't a monarchy at all, it's an egalitarian society. There is space for all of them and they all supplement each other. We can't reduce them to a single fundamental science and we don't need to. The relation between them is not linear but convergent.

THE VANISHING SOUL

We badly need to be clear about this point today because the stopgap device that used to obscure the need for it is vanishing. When Descartes first introduced this model, it notoriously had another component which was supposed to take care of the first-person

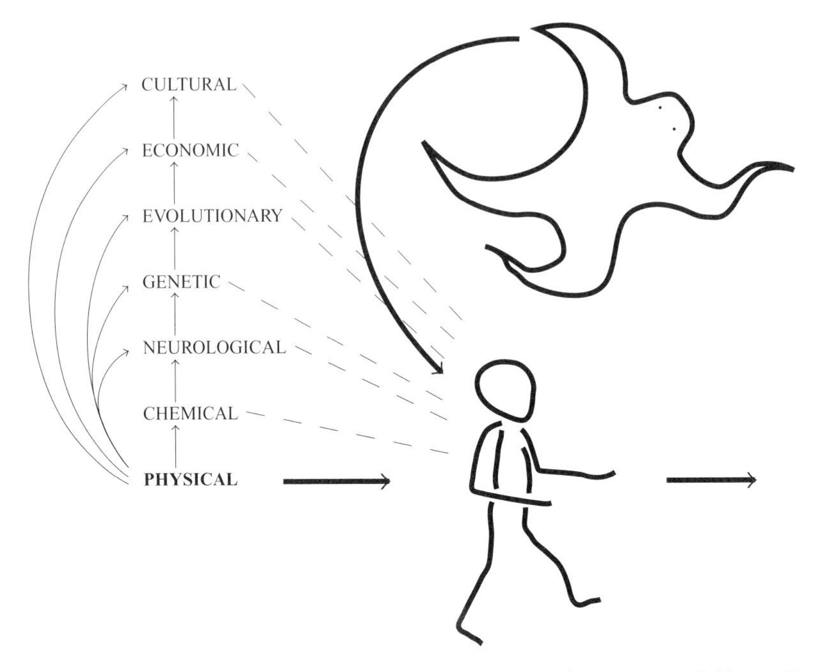

FIGURE 1.2. Causation: seventeenth-century reductionist model (+ soul).

viewpoint – the immortal soul, the seat of consciousness (Figure 1.2). That soul was still an accepted part of the model in Newton's day. But it has always been an unsatisfactory device. It was far too simple to deal with the manifold functions of consciousness, and far too disconnected from the physical mechanisms to be capable of driving them. So it was gradually sidelined.

I suspect that it is this kind of soul that Dawkins, Blakemore and their allies attack. Quite rightly, they insist that a brain doesn't need this extra, disembodied entity to drive it. Brains work because they are parts of living bodies. But then, our ordinary notion of the active self isn't the notion of such a disembodied soul either. It's a notion of the whole person – not divided into separate body and mind – of whom the brain is just one working part.

The disembodied soul was certainly unhelpful and we do not now invoke it. *But without it, the rest of the seventeenth-century*

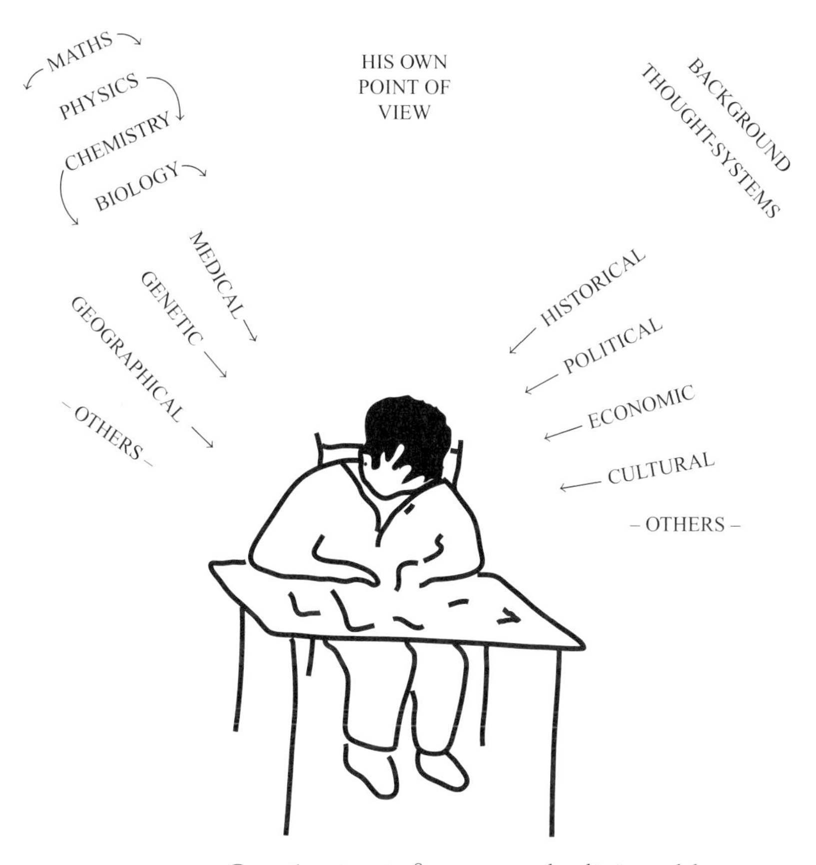

FIGURE 1.3. Causation: twenty first-century pluralistic model.

model doesn't really make sense. We need a new one that does justice to the many different kinds of question that we ask and the ways in which they all converge. Accordingly, let's look at Figure 1.3.

EXPLAINING THE THINKER

This is a picture of somebody working on a really hard problem. (He happens to be male and I have found it impossible to keep using two sets of pronouns throughout the discussion of his predicament, but I hope readers will be able to fill these in for themselves.) This person might be Darwin or Einstein or Jane Austen or Hildegard of Bingen, he might be Napoleon or Boadicea planning a campaign, he might be the chairperson of the Mafia organising a heist, he might be someone busy on an article for the *Journal of Consciousness Studies* or working on any of the difficult choices that Blakemore has listed. The details of the problem don't matter. What does matter is (1) that it is a hard one, hard enough to need careful attention, (2) that it is something on which he will eventually need to act, so that our question about the nature of action will finally be relevant. On ordinary assumptions, what is decided here will determine later action and thereby affect the outside world. I shall suggest that these assumptions are entirely correct.

As this person sits and thinks, we can imagine the converging, but not necessarily competing, lines of explanation as raying out from him or her on all sides. The arrows in this new diagram don't represent forces charging in to 'drive' the thinker like a passive vehicle. Instead, they are lines of sight for the observer – viewpoints – angles from which we can look at a most complex process. They are positions that we might take up if we – as outside observers – want to understand his thinking.

If we are taking that thinking seriously, the first thing that we shall try to do is to grasp the position that I have put at the top – namely his own point of view. We try to follow his reasoning as he himself understands it. If it then seems satisfactory, we may simply accept it as our own. But if it does not, we will see reason to move to one of the other points of view, so as to find out what else is needed.

In the crudest case, if the ideas involved seem really crazy, we look at the medical angle and see whether the thinker himself may actually be ill. (Perhaps he might have a brain tumour.) We can also, if we see the need, ask questions about his background, about a wide range of factors that may have influenced him. (I have grouped these factors roughly on the left of the drawing.) But this kind of supplementation is not normally appropriate unless we have reason to think that the views themselves are faulty. Before resorting to it, we bring in various conceptual schemes simply to fill in gaps that we find in his thought, to extend it and to see whether we can make better sense of it. I have put a few of these conceptual schemes on the right.

They are ones that bear on the subject matter that he is dealing with, rather than on his own peculiarities.

Thus, in Napoleon's case, economic historians might find themselves at odds with political and military ones because they use different abstractions to concentrate on different aspects of his problem. In a case like Darwin's, there is enormous scope for these conflicts because his ideas are so wide-ranging that they raise questions for a whole gamut of disciplines, offering a corresponding number of opportunities for clashes. For instance, in his own day his biological suggestions seemed to conflict with important doctrines of physics, since (as Lord Kelvin pointed out) it did not seem possible that an earth as old as the one that Darwin envisaged could possibly have kept its heat long enough to allow the development of life. The calculations from physics overlooked the heat input from the spontaneous decay of natural radioactive substances which was not appreciated at the time. *My analogy with the relation between different maps is meant to draw attention to just this kind of clash between conceptual schemes.* Though there is indeed only one world, the various disciplines necessarily describe it differently by abstracting different patterns from it. While they remain unaware of each other, they are liable to commit themselves to views that turn out to conflict. And when this is noticed, both parties need to work to make their conclusions somehow compatible. In Darwin's case it was physics that was wrong, but very often changes are needed on both sides. Making those changes does not, however, mean getting rid of their different methods so that they end up with a single pattern. Nor does it mean that one discipline will eliminate the other. They continue to present different pictures, like the different maps of the world, but now with a better understanding of how they should be related.

'FUNDAMENTAL'?

In the case of our worried thinker, then, no one of the enquiries that we can make is going to give us a 'complete explanation' of his thought. Indeed, it is not clear what a complete explanation would be, since there are an infinite number of questions that might be asked about it. Each enquiry necessarily shows only one aspect of what is going on. If we want a fuller view of what's happening, we will have to put a number of them together. Is there any reason to expect that one of these kinds of explanation should be more fundamental than the rest? Is any such hierarchy necessary? In particular, is there any reason, when we talk about action, to prioritise facts about the brain over other explanatory facts?

It is not clear why such a hierarchy should be needed. When we pick out one explanation as being 'fundamental', we normally mean that it is specially relevant to the particular enquiry that we want to make. If indeed we are neurologists and our enquiry is about whether (for instance) Einstein's brain was really different from other people's, then we might reasonably call the details of that brain fundamental. But for most other purposes we simply take it for granted that a well-functioning brain is needed as a background condition of all thought – as is also the rest of the nervous system and the other organs – and we assume that, so long as it functions properly, the details of its working do not matter.

This assumption does not reflect any underestimation of neurology. It is an assumption that we have to make because thought does not remain a private matter. It leads to speech and to other actions in the world, and these must be intelligible to other people besides their originator. The ideas that a thinker comes up with must be ones that these others can understand too, not ones that would only fit a brain that is physically like his or her own. Original thinkers, in explaining their ideas to others, do not proceed by sending out diagrams of their own brain-states, but by speech and actions in the public world. Thus, no amount of information about Einstein's brain would enable a neurologist who was ignorant of physics to learn anything from that brain about relativity theory.

WHY PLURALISM IS NEEDED

I have been making a case for cognitive pluralism – for the view that all explanation, and particularly the explanation of human action, quite properly proceeds by many non-competing but convergent methods, because it involves answering questions that arise from different backgrounds. That is why explanation cannot be reduced to a single fundamental method.

I would like now to return to Colin Blakemore's alternative formulation and to note the difficulties raised by its more reductive approach – difficulties that I think show why something pluralistic is needed. The trouble here begins with the word 'alone' in his first sentence. ('The human brain is a machine that alone accounts for all our actions.') Certainly a suitable brain is needed as one causal factor in all human action. But how could it be considered as the only one?

As just noticed, if we want to account for somebody's action – that is, to explain it – the first thing that we need to know about will be his or her own point of view, beliefs, skills and conceptual schemes, motives, background. Without understanding where he or she starts we can make no progress. After that, we shall also need to know more about the options open to him or her, which means examining the social and physical life around the individual.

The 'causal chain that runs back to the origin of all life and down to the nature of atoms and molecules' that Blakemore mentions is not really just one chain passing through this agent's brain. It is a network that runs crossways through every aspect of his or her life and much of the surrounding world. If, when we have investigated it, we still find his or her action unintelligible, we may then start to enquire about the state of the agent's body, including the brain, to see whether some illness is distorting judgement. But normally, the explanation of actions goes on successfully without any investigation of the brain-state at all.

It should be noted that the neurological kind of reduction is not the only one available. It would be equally possible (and equally misleading) to say that the realm of background thought in the world determines the whole thing. For instance, it could be argued that Einstein's next move was fully determined by the state of physics in his day – by the totality of moves made by previous physicists, which left only one path open. That intellectual realm would then provide its true explanation. This is the kind of suggestion that is put forward by reductionists of a different stripe, those who want to interpret all phenomena in terms of patterns of information. And of course, with hindsight, explanations of this sort are often useful. But they too obviously depend on illicit abstraction – on picking out a single favoured pattern as sovereign and neglecting the rest of the world.

What, then, does this mysterious word 'alone' mean? It is my impression that it is really intended to negate only one other possible cause – namely, conscious thought, first-person activity, which is ruled out as a causal factor in producing action because it is believed to involve a detached Cartesian soul.

This seems clear when Blakemore writes, 'It seems to me to make no sense (in scientific terms) to try to distinguish sharply between acts that result from conscious intention and those that are pure reflexes or are caused by disease or damage to the brain.'

It is surely rather strange to dismiss this distinction as unscientific, since it is one that a consultant neurologist who was examining a patient would undoubtedly think important, and indeed one that is relevant in many other areas of medicine. Certainly the distinction is not always *sharp*. There are borderline areas. But to suggest that it does not arise at all – that there are no clearly distinguishable 'acts that result from conscious intention' – is to suggest that, for all we know, the writing of *On the Origin of Species* might not have a consciously intended act of this kind but just an inadvertent spasm, comparable to a reflex. And this is really not convincing.

CONCLUSION: SUBJECTIVITY IS AN OBJECTIVE FACT

The philosophical conclusion that I want to stress here is that conscious thought has a legitimate and essential place among the causal factors that work in the world. It is not a spooky extra but a natural process. In a species such as ours, it is an integral part of normal behaviour. Descartes was wrong to export it to a metaphysical ghetto. Our inner experience is as real as stones or electrons and as ordinary an activity for a social mammal as digestion or the circulation of the blood. The capacity to have this conscious experience, and to use it effectively in making choices, is one that has evolved in us, and in many other species, just as normally as our capacities to see, hear and walk.

We need to notice that such a capacity could not possibly have evolved – as Blakemore suggests – merely as an idle epiphenomenon, surface froth, a shadow-show with no effect in the real world. Natural selection can only work on real consequences. It can therefore only promote things that are effective. There is no way in which it could have got a grip on such an ineffectual shadow and made conscious thought a normal accomplishment in the species – which it certainly now is. The reason why this power of practical thinking has been able to evolve is that it is a useful way of producing well-judged action.

This conclusion does not, then, involve any extravagant metaphysics. When we say that someone acts freely, deliberately and responsibly, this does *not* mean that a separate soul does so, cut off from the influences around it. It simply means that *he or she does this action as a whole person,* attending to it and being well aware of what they are doing – not (for instance) absent-mindedly or madly or under some outside pressure such as hypnosis. Of course this agent needs to have a brain – and no doubt some genes – in good order to make this choice. But it is he or she, the whole person, who uses that brain, just as they use their legs to walk and their eyes and hand in writing.

2 The definition of human nature

MERLIN W. DONALD

Our definition of human nature gives us a conceptual foundation for our ideas about human rights, individual responsibility, and personal freedom. These ideas were originally derived from the liberal humanities, and are ultimately the secular modern descendants of the concept of a 'natural law' based on earlier religious and philosophical traditions. In this context, this is not a trivial exercise. It provides a conceptual foundation for our legal system, as well as our constitutional protections of human rights. Since the time of Charles Darwin there have been many attempts to define human nature in more scientific terms. In effect, this has amounted to an attempt to derive a new kind of natural law, based largely on scientific evidence, and especially on the theory of evolution. Here I am not speaking of Social Darwinism, an earlier intellectual movement that naively tried to extrapolate the laws of natural selection to human society, but of more recent empirical attempts to construct a culturally universal description of the human condition, and to explain it in terms of evolution and genetics.

In such attempts, human nature is usually defined as having been fixed when our species evolved in the Upper Palaeolithic, and this suggests that we have been genetically engineered to survive under the special conditions of late Stone Age civilisation. This raises the disconcerting possibility that human nature might prove to be maladaptive in today's high-tech, fast-moving, urbanised world. On the other hand, the logic leading to this conclusion is not compelling. It is based on two assumptions. The first is that human beings have a fairly rigid set of constraints on their mental and social life, imposed

The New Brain Sciences: Perils and Prospects, ed. D. Rees and S. Rose.
Published by Cambridge University Press.

by an inflexible genetic inheritance. The second is that human mental and social life is determined largely by organismic variables, and that the human mind can be treated like that of any other species.

But there is an alternative view of human nature, based on scientific evidence and evolutionary theory that comes to a different conclusion, and fits existing scientific evidence better. It is based on a different set of assumptions. The first is that human nature has been characterised by its flexibility, not its rigidity. This is due largely to the overdevelopment of conscious processing, and those parts of the brain that support it. The second is that human beings, as a species, have evolved a completely novel way of carrying out cognitive activity: distributed cognitive–cultural networks. The human mind has evolved a symbiosis that links brain development to cognitive networks whose properties can change radically. Critical mental capabilities, such as language and symbol-based thinking (as in mathematics) are made possible only by evolving distributed systems. Culture itself has network properties not found in individual brains. The individual mind is thus a hybrid product, partly organismic in origin, and partly ecological, shaped by a distributed network whose properties are changing. Our scientific definition of human nature must reflect this fact, and free itself, not only of pre-scientific notions about human origins, but also of restrictive and antiquated notions about organismic evolution.

One consequence of this idea is that 'human nature', viewed in the context of evolution, is marked especially by its flexibility, malleability and capacity for change. The fate of the human mind, and thus human nature itself, is interlinked with its changing cultures and technologies. We have evolved into the cognitive chameleons of the universe. We have plastic, highly conscious nervous systems, whose capacities allow us to adapt rapidly to the intricate cognitive challenges of our changing cognitive ecology. As we have moved from oral cultures, to primitive writing systems, to high-speed computers, the human brain itself has remained unchanged in its basic properties, but has been affected deeply in the way it deploys its resources. It develops in a rapidly changing cultural environment that is largely of its own making. The result is a species whose nature is unlike any other on this planet, and whose destination is ultimately unpredictable.

HUMAN ORIGINS

Anatomical and DNA evidence suggests that we emerged as a new species about 160 000 years ago in Africa, and then migrated over most of the Old World, replacing all other existing hominids. The human adventure had its starting point with the Miocene apes of Africa, which existed about 5 million years ago, and were our common ancestors with modern chimpanzees and bonobos. The primate line that led to humanity diverged from the orang-utans about 11 million years ago, from gorillas about 7 million years ago, and from a species similar to modern chimpanzees about 5 million years ago. The next phase of the human story proceeded through a series of intervening human-like or hominid species, which provide a well-documented morphological linkage to our distant ape cousins. The first to diverge in the human direction were the australopithecines, who dominated the hominid niche from 4 million until about 2 million years ago. This species crossed a major hominid threshold, in becoming the first primate species to walk erect. However, their brains did not increase greatly in volume from those of their ape-like predecessors. Although this period saw the emergence of some important human traits, such as pair-bonding, there is no good evidence to suggest that australopithecines made any major progress toward our distinctive cognitive capacities.

The first move in that direction came much later, about 2 million years ago, with the first member of the genus *Homo*, *Homo habilis*. Habilines had brains that were slightly larger than those of their predecessors and had a characteristically human surface morphology. This change was the result of an expansion of tertiary parietal cortex, and remains characteristic of modern humans. Habilines were replaced, only a few hundred thousand years later, by larger hominids that were much more human-like in appearance, and had a much higher encephalisation, or brain–body ratio, which eventually reached about 70% of that of modern humans. These archaic hominids, variously called *Homo erectus*, archaic *Homo*, or pre-sapient *Homo*, had larger brains gained at the expense of a reduced gut size, a metabolic trade-off necessary to service their energy-expensive brains. This means that they could not have scavenged for a living, and probably had to pre-process foods, such as root vegetables and meat, before such foods would have been digestible. This demanded a considerable improvement in cognitive capacity, since pre-processing food requires foresight, and the tools needed for this required the diffusion of complex skills to all members of the hominid group. Very soon after appearing in Africa, *Homo erectus* migrated to Eurasia, and over the next few hundred thousand years began to domesticate fire, eventually mastering the considerable technical demands of its continuous use. *Homo erectus* also developed better stone tools (the so-called Acheulian tool culture, which originated with them, continued in use for well over a million years), and underwent a series of other major changes, including the adoption of progressively more challenging campsites, which were sometimes far removed from the sources of toolmaking materials and water. This implies changes in their strategies for finding and transporting food and water, their division of labour, and their diet and hunting capabilities. I have referred to this long period as the first transition; that is, the first period during which hominids took major steps towards the development of the modern form of the human mind.

The second transition occurred much later, starting about 500 000 years ago, and ended with the appearance of our particular species. This period was marked by a pattern of rapid brain expansion that continued until relatively recently. One of its prominent features was the appearance of the Mousterian culture, which lasted from about 200 000 to about 75 000 years ago, and had such distinct features as the use of simple grave burials, built shelters and systematically constructed hearths. The Mousterian toolkit was more advanced, and included more differentiated types of tools, and better finish. These two transitions culminated in our own speciation, and might be taken as the end of the human story, from a biological standpoint.

But from a cognitive standpoint, it was not the end. I have proposed a third transition, one that was largely culturally driven, and did not involve much, if any, further biological evolution. Nevertheless, the cognitive grounds for postulating this third transition are identical to my reasons for postulating the first two: fundamental changes in the nature of mental representation. This third period of change, which started about 40 000 years ago and is still under way, was characterised by the invention and proliferation of external memory devices. These included symbols, and the use of symbolic artefacts, including scientific instruments, writing and mathematical systems, and a variety of other memory devices, such as computers. Its greatest product was the growth of cognitive networks of increasing sophistication. These networks, which include governments, and institutionalised science and engineering, perform cognitive work in a different way from individual minds. This has produced a cognitive revolution that is easily a match for the two previous ones.

These periods of transition form a chronological backbone for any scenario of cognitive evolution, but do not constitute the primary evidence for a deeper cognitive theory of origins. The most important evidence for the latter comes from cognitive science and neuroscience. One of the most controversial issues revolving around human evolution is the question of whether humans have specialist or generalist minds. There is some evidence to support both positions. On the one hand, we have some unique intellectual skills, such as language and manual dexterity, that seem to call for a specialised brain. Such skills are defined as innate or instinctual by many theorists, because they are found universally in all cultures, and are difficult or impossible to change. They are regarded as an integral part of human nature, built right into the genome, and the result of adaptation to environmental pressures that were specific to the time of our speciation.

On the other hand, the human mind is notable for its flexibility and general-purpose adaptability. Our abstract thinking skills can be applied in a number of domains, such as mathematics or polyphonic music, that did not exist at the time we evolved. Thus, our proficiencies in these areas could not be the result of the direct evolution of specialised abilities in maths and music, since natural selection can only act upon existing alternatives. The evidence we have on the human brain seems to rule out the strong form of the notion that the power of the human mind is the result of many specialised neural adaptations. We have many examples of specialisation in other species, and we know what such adaptations look like. Such cognitive abilities usually reside in identifiable neural structures, or modules. For instance, birdsong resides in certain clearly delineated brain nuclei that are uniquely developed in songbirds. There are dedicated neural modules of this sort in many species, supporting such capacities as echolocation in bats, magnetic navigation in migrating birds, and stereo vision in primates, and they all depend upon specialised sensory organs, with associated brain structures. Humans have the specialist profile common to primates, including excellent stereo vision, but, with the exception of our vocal tract, we have no modules that are unique to us. Some thinkers seize upon spoken language as a unique specialist adaptation, and, to a degree, it is. But the catch is that we do not need our vocal tract to produce language. Sign language, as spoken by the deaf, makes no use of our special vocal apparatus. Moreover, although it is normally localised on the left side of the brain, language does not seem to be restricted to any particular lobe, or subsection, of the association cortex, or even to the left side. Thus, it is difficult to claim that there is a neural module for it.

If we review the deep history of the vertebrate brain, we find that very few new modules have evolved during its 500 million years of evolution, and that each has taken a great deal of time. For example, the motor brain underwent a major modular reorganisation after vertebrates moved onto land, and the standard motor brain architecture of land animals, particularly that of the higher mammals, is quite different from that of aquatic species. However, the aquatic brain is retained, virtually intact, within our nervous system, although it has been surrounded by powerful new modules. The aquatic brain evolved to control the wriggling movements, and alternating rhythms, typical of fishes. But a complex web of other structures has evolved to support the many new demands faced by land vertebrates. The neuropsychological study of the component structure of the human brain has not turned up any truly novel structures, in the form of lobes, nuclei or ganglia that do not have an equivalent homologue in apes, or any new transmitter systems, neurotubular networks, or even dendritic structures that are completely unique to our species. Admittedly, the human brain's enlargement is unprecedented, and there have been some epigenetic changes in synaptic growth patterns and interconnections, but the basic architecture of the primate brain was not changed by this expansion. The most dramatic change in connectivity is in the prefrontal cortex (see Deacon, 1996), where a disproportionate increase has caused the frontal cortex to split into subregions and invade areas that it does not typically infiltrate in other species. Our frontal lobes have more influence, and our species is highly frontalised in its intellectual strategies. Moreover, the association zones of our neocortex, cerebellum and hippocampus are much larger than corresponding structures in chimpanzees. But even this trend is not unique to us. The increased encephalisation of primates has been ongoing for tens of millions of years, and roughly the same structures have continued to increase in our own species. Just as there was a tripling of brain size between primitive monkeys and chimpanzees, there was a tripling between apes and us. This presents us with several questions. If humans inherited a brain architecture that is relatively universal in the higher vertebrates, how is it that our minds are so special? How could we have evolved capacities for language, art and technology without much change at the modular level in the nervous system, given that less drastic cognitive changes in many other species have required new modules?

One answer is to remember that brain evolution cannot explain everything about us. Culture is a huge factor in human life, and the distributed networks of culture have been accumulating knowledge since very long ago. The presence of culture as an immense collective cognitive resource is a novel development in evolution. Culture is constrained by biology, of course, and there is often a delay between evolutionary changes in the brain and major cultural advances. For example, improved tool use and the domestication of fire came many generations after the period (about 2 million years ago) when there was rapid hominid brain evolution. These large time lags between neural and cultural change suggest that the increase in brain size was not initially driven by immediate improvements in toolmaking and fire-tending; they had not yet happened. The human brain evolved most of its physical features for other reasons, related to such things as diet, habitat, social coordination and developmental plasticity. Our major cultural achievements have evidently been the delayed by-products of biological adaptations for something else. The same is true of the second great hominid brain expansion which accompanied the emergence of our modern vocal apparatus. Our brains reached their modern form less than 200 000 years ago, but evidence of rapid cultural change only appeared around 50 000 years ago. Our larger brain and vocal tract must have been the products of other significant fitness gains, perhaps related to survival during the Ice Ages, which did not cause an immediate revolution on the cultural level. Vocal language could have been the major change, and would not have left any immediate evidence of its arrival. The long-term result was increased cognitive plasticity. The implication is that archaic humans evolved a set of flexible intellectual skills that led to gradual and continuous cultural change.

CROSSING THE RUBICON

The key to rapid cultural change is language. Humans are the only minds in nature that have invented the symbols and grammars of language. Some apes can be trained to use human symbols in a limited way, but they have never invented them in the wild. This even applies to enculturated apes who have acquired some symbolic skills, and we conclude from this that the mere possession of symbols cannot in itself bring about radical change. It is the capacity to create symbols that they lack. Crossing the abyss between pre-symbolic and symbol-driven cognition was a uniquely human adventure, and consequently there is a huge gap between human culture and the rest of the animal kingdom. Any comprehensive theory of human cognitive evolution ultimately stands or falls on its hypothesis about how this gap came to be bridged.

Cognitive science is broadly divided between the artificial intelligence, or AI, tradition, which builds symbol-driven models of mind, and the neural net tradition, which develops models of simulated nervous systems that learn without using symbols, by building hologram-like memories of experience. A neural net is basically a tabula rasa network of randomly interconnected memory units, which learns from environmental feedback, by building associations in a relatively unstructured memory network, much the way many animals do. Artificial intelligence models, on the other hand, depend on preordained symbolic tools – in the form of elementary categories and rules – given to them by a programmer, and these are used to construct symbolic descriptions of the world, rather like those that humans build with language. But there is a crucial difference between such artificial expert systems and the human mind. Expert systems have no independent knowledge of the world, and remain locked in at the symbolic level, so that to understand a sentence, they are limited to looking up symbols in a kind of computational dictionary, each definition pointing only to more words or actions, in endless circles of dictionary entries. In such a system, there is no path back to a model of the real world, and symbols can only be understood in terms of other symbols. Since, as Wittgenstein observed, the vast majority of words cannot be adequately defined with other words, this is not a trivial limitation. The development of the AI tradition has run into a brick wall, as Dreyfus predicted twenty years ago, precisely because it cannot cross the boundary-line between pre-symbolic and symbolic representation, and access the holistic, non-symbolic kinds of knowledge that humans use to inform their symbolic constructs.

The key question of human cognitive evolution might be rephrased in terms of this dichotomy: somewhere in human evolution the evolving mammalian nervous system must have acquired the mechanisms needed for symbol-based thought, while retaining its original knowledge base. To extend the metaphor, it is as if the evolving mammalian mind enriched its archaic neural net strategy by inventing various symbol-based devices for representing reality. This is presumably why the human brain does not suffer from the limitations of AI; it has kept the basic primate knowledge systems, while inventing more powerful ones to serve some non-symbolic representational agenda. But, how could the evolving primate nervous systems of early hominids have crossed the pre-symbolic gap? What are the necessary cognitive antecedents of symbolic invention? Cognition in humans is a collective product. The isolated brain does not come up with external symbols. Human brains collectively invent symbols in a creative and dynamic process. This raises another important question: how are symbols invented? I attribute this ability to executive skills that created a nervous system that invented representation out of necessity.

When considering the origins of a radical change in human cognitive skill, we must look at the sequence of cultural changes, including the cultures of apes and hominids. The cognitive culture of apes can be epitomised by the term 'episodic'. Their lives are lived entirely in the present as a series of concrete episodes, and the highest element of memory representation is at the level of event representation. Animals cannot gain voluntary access to their own memory banks, because, like neural nets, they depend on the environment for memory access. They are creatures of conditioning, and think only in terms of reacting to the present or immediately past environment (this includes even their use of trainer-provided symbols, which is very concrete). Humans alone have self-initiated access to memory. This may be called autocueing, or the ability to voluntarily recall specific memory items independently of the environment. Consider an animal moving through a forest; its behaviour is defined by the external environment, and it can be very clever in dealing with that environment. But humans can move through the same forest thinking about something totally unrelated to the immediate environment – for instance the recent election, a movie or an item in the newspaper. In thinking about some topic, the thinker pulls an item out of memory, reflects on it, accesses another memory item, connects this to the previous idea, and so on, in recurrent loops. This reflective skill depends on voluntary autocueing; each memory item is sought out, precisely located and retrieved, preferably without retrieving a batch of other unwanted items, and without relying on the environment to come up with the relevant cues to help find the item. Our ability to transcend the immediate environment could not have developed without autocueing skill. Note that I am not saying we can introspect on the process by which we voluntarily access memory. We do not have to be aware of the retrieval process to have voluntary control over it. Language is 'voluntary' cognition, but we have no awareness of where the words are coming from when we speak. The first symbolic memory representations had to gain explicit access to the implicit knowledge latent in neural nets. The initial adaptive value of the representational inventions of early humans would have been their ability to provide retrieval paths to a knowledge base that was already present, but not voluntarily accessible, in the primate brain. But, given the functional arrangement of the primate brain, where would such paths have been built?

THE FIRST STEP TO LANGUAGE: MIMESIS

The first cognitive transition occurred between 2.2 and 1.5 million years ago, when major changes in the human genome culminated in the appearance of *Homo erectus*, whose achievements indicate some form of improved memory capacity. This species produced (and used) sophisticated stone tools, devised long-distance hunting strategies, including the construction of seasonal base camps, and migrated out of Africa over much of the Eurasian land mass, adapting to a wide variety of environments.

Many evolutionary theorists are fixed on the idea that there was only one great cognitive breakthrough for humans: language, that this breakthrough came early, with *Homo erectus*, and that all higher human mental abilities followed from it. Bickerton (1990) argued that some form of proto-language must have existed at the time of *Homo erectus*, which might explain early hominid cultural achievements with a single adaptation – a sort of grammarless language – that later evolved into modern speech capacity. Pinker (1994) has suggested that grammar itself started its evolution early, and that some parts of a language module must have already been in place in *Homo erectus*.

I find this unconvincing. First, archaeological evidence doesn't place speech so early in evolution; neither of the principal markers for human language – the descended larynx and rapid cultural change – appears in the archaeological record until *Homo sapiens*, who evolved more than a million years later. Second, early hominids had no existing linguistic environment, and even proto-language would have required a capacity for lexical invention. This issue is crucial, because it raises the question of the autocueing of memory: lexical inventions must be self-retrievable, that is, autocueable. True linguistic symbols, even the simplest, could not suddenly pop up in evolution before there was some principle of voluntary memory retrieval in the hominid brain; to be useful, lexical inventions had to be voluntarily retrievable and modifiable, as well as truly representational acts, intentionally modelling some aspect of reality.

Before lexical invention became a realistic possibility, it was necessary to establish voluntary retrieval, or autocueing, in the pre-linguistic brain. The same adaptation would also have provided the cognitive prerequisite for a number of non-verbal representational skills. After all, language is not the only uniquely human cognitive advantage that has to be explained in evolution (Premack, 1986). If all our higher thought-skills were based on our linguistic capacity, how could we account for the virtual autonomy of some non-verbal forms of human intelligence? A good evolutionary theory of pre-linguistic adaptation should try to account for as many of these skills as possible, while providing the cognitive grounds for language.

My key proposal is that the first breakthrough in our cognitive evolution was a radical improvement in voluntary motor control that provided a new means of representing reality. *Homo erectus's* gift to humanity was mimetic skill, a revolutionary improvement in voluntary motor control, leading to our uniquely human talent for using the whole body as a subtle communication device. This body skill was mimesis, or a talent for action-metaphor. This talent, without language, could have supported a culture that, in terms of its tool-making abilities, was much more powerful refinements of skill, and flexible social organisation, than any known ape culture.

Mimetic skill logically precedes language, and remains independent of truly linguistic modes of representation. It is *the* basic human thought-skill, without which there would not have been the evolutionary opportunity to evolve language. Mimesis is an intermediate layer of knowledge and culture, and the first evolutionary link between the pre-symbolic knowledge systems of animals and the symbolic systems of modern humans. It is based in a memory system that can rehearse and refine movement voluntarily and systematically, in terms of a coherent perceptual model of the body in the surrounding environment, and is based on an abstract model of models that allows any action of the body to be stopped, replayed and edited, under conscious control. This is inherently an autocueing route, since the product of the model is an implementable self-image. Although the precise physiological mechanism of this system is not known, its functional retrieval path employs kinematic imagery. The principle of retrievability was thus first established at the top end of the motor system; and retrievable body-memories were the first true representations.

Mimesis is a supramodal skill. A mimetic scenario can be acted out with eyes, hands, feet, posture, locomotion, facial expression, voice, or any other motor modality, or combination of modalities. This is evident in the uniquely human behaviour pattern known as rhythm, which is the motor translation of an abstract sound pattern, or the conversion of sound into motion. Rhythm is truly supramodal: revellers at a rock concert use every muscle in their bodies to convert an abstract sound pattern into movement. But the more complex human motor skills necessitate more than a capacity for supramodal rehearsal: they also require a capacity for purposive sequencing of large-scale patterns of action over longer periods of time, such as those used in advanced toolmaking. This assumes a larger self-modelling capacity whereby a series of actions can be imagined and then altered or re-sequenced. This kind of extended kinematic imagination is still the basis of human non-verbal imagination, and is essential to the training of those who work with the body, such as actors and gymnasts. Although it is sometimes seen as primarily visual, non-verbal imagination is a body-based skill that captures visual images in its wake. It is not an accident that the ancient mnemonic method favoured by the Greeks and later European cultures did not rely on static visual imagery, but rather on generating an image of self-motion inside an imaginary visual space, where the kinematic image was made the engine of visual recall.

The universality of these uniquely human body skills is still evident in children of all cultures, who routinely practise and refine their motor skills without training or conditioning; images of boys bouncing a ball off the wall over and over again, or girls skipping a rope endlessly, come to mind. An advance in human motor representation of this magnitude would automatically have had ramifications in the area of expressive capacity. Actions and events could be represented and re-enacted independently of the environment; and this resulted in improved toolmaking and tool use, and in constructional and other instrumental skills. But, as in many evolutionary adaptations, mimetic skill would have had unforeseen consequences: now hominids had a means of re-presenting reality to themselves and others, by the use of voluntary action. This means that hominids could do much more than rehearse and refine existing movement patterns; they could also imagine and invent completely new ones, as human gymnasts, dancers, actors and divers still do. And they could re-enact events and scenarios, creating a sort of gestural proto-theatre of everyday life. The body became a tool for expression; it was just a matter of discovering the social utility of this possibility.

The expressive and social aspect of human mimetic skill may be called pure mimesis. For a long time (more than a million years) hominids subsisted on a mimetic culture based on improved voluntary motor skill, extensive use of imitation for pedagogy and a much more sophisticated range of voluntary facial and vocal expressions, along with public action-metaphor, which formed the basis of most custom and ritual. Could such a language-less culture have carried *Homo erectus* to the heights he achieved? There is strong support for the power and autonomy of non-verbal mind in the study of modern humans. One line of evidence is the enduring cultural autonomy of mimesis. Whole areas of modern human culture continue to function magnificently with a minimal use of language. These include the practice and teaching of many crafts and occupations: games, especially children's games; many aspects of custom, social ritual and complex interactive scenarios such as those documented by Eibl-Eibesfeldt (1989) and others (Argyle, 1975); athletic skill; and many group expressive customs – for instance, the systematic use of group laughter as a means of ostracism or punishment, and culture-specific customs for indicating deference, affection, manliness, femininity, tolerating pain, celebrating victory, maintaining group solidarity, and so on. These aspects of culture do not depend on language skill, either in their original invention, or in their transmission from one generation to the next.

Another line of evidence is neurobiological. These areas of skill are typically resilient in certain cases of global aphasia. This is especially clear in temporary aphasias caused by some types of epilepsy, where patients may lose all use of language (including inner speech) for a few hours, but remain conscious and able to function on a non-symbolic level. They can still find their way around in a purposive manner, operate a relatively complex device like a radio or elevator, and manage mimetic social communication (for instance, they know when they are having a seizure and can communicate this to others by gesture). This implies that mimetic skills come from an autonomous level of representation in the brain, unaffected by temporary but complete loss of language.

Further evidence for the independence of pure mimesis comes from the documented lives of illiterate deaf people from the eighteenth and nineteenth centuries, before the diffusion of formal sign languages. Without any training to help them communicate, such individuals had to survive without the lexical, syntactic or morphological features of language. They could not hear, and thus did not have a sound-based lexicon of words; they lacked an oral lexicon; they could not read or write and thus lacked a visually based lexicon; and in the absence of a deaf community with a formal sign language, they had no signing lexicon. None of the lexical components of language was available, and this would have eliminated the possibility of constructing anything we might recognise as true linguistic representations. Yet, they often lived remarkable lives (Lane, 1984), and by recorded accounts were quite sophisticated in their use of pure mimesis, both in constructional skill and in communicative and metaphoric gesture.

Mimetic representation is an autonomous, uniquely human level of mind that still supports the non-linguistic cognitive infrastructure of human society. It allowed humans to break the primate mould and construct retrievable memory representations for the first time. It also led to a slow-moving process of cultural change that culminated in the distinctively human cultures of late *Homo erectus*, and set the stage for a second drastic innovation that would create a much more powerful representational device.

THE SECOND STEP: MYTHIC CULTURE

The second transition, from mimetic to mythic culture, was made possible by language. As a result, the scattered, concrete repertoire of mimetic culture came under the governance of narrative thought and ultimately, integrative myth. Archaeological markers indicate that a long transition period, from 500 000 to 100 000 years before present, preceded the appearance of modern *Homo sapiens*. This is the period when language is most likely to have evolved. Language involves a different type of cognitive operation from the holistic motor strategy underlying pure mimesis. It depends primarily upon a capacity for inventing and retrieving thousands of lexical items – words – along with the rules that govern their use, and constructing narrative commentaries out of these lexical items. Words were the first true symbols, and language in this sense is the signature of our modern human species. Evolutionary pressures favouring such a powerful representational device would have been much greater once a mimetic communicative environment reached a critical degree of complexity. Mimesis is inherently an ambiguous way of representing reality, and words are an effective means of disambiguating mimetic messages. Modern children still acquire speech in this way, with most of their early utterances embedded in mimetic exchanges, such as pointing, tugging, prosodic voice sounds, eye contact, non-linguistic sounds and gestures, and mimetic whole body movement. Even when the young child is talking to itself, it is usually in a mimetic context.

Lexical invention is a constant process of labelling, defining and differentiating aspects of the perceived world (including the products of speech itself). Humans are constantly inventing new lexical items or acquiring them from others, and oral languages are seldom static for long. This reveals a continuing tension between lexical inventions and their significations, as if there was a natural tendency for the system to keep differentiating and defining reality. Like mimesis, language is at core a thought-skill, but rather than using the holistic, quasi-perceptual strategy of mimetic motor skill, it employs true symbols and constructs narrative descriptions of reality.

Spoken language provided humans with a second form of retrievable knowledge and a much more powerful way to format their knowledge. The natural product of language is narrative thought, or storytelling. Storytelling had a forerunner in mimetic event re-enactment, but is very different in the means by which it achieves its goal, and much more flexible in what it can express. Mimetic re-enactment is bound to imagery of the original event being depicted, but the quintessential narrative act – verbally labelling agents, actions and their relationships – lifts the observer outside of space and time, allowing the component parts of the story to be examined, reassembled and shared much more freely.

Spoken language altered human culture not merely in the number and complexity of available words and grammars, but in the shared products of oral cultures. The collective use of narrative thought led inevitably to standardised descriptions: shared, agreed-upon versions of past events. These formed the basis of myth and religion, which were the direct products of evolving linguistic skill. It is telling that mythic invention seems to have preceded any further advances in human toolmaking. Even the most technologically primitive cultures have fully developed oral languages and mythic systems. However, the new oral cultures did not abandon mimetic representation; to the contrary, they encompassed the more concrete, pragmatic culture of mimesis, which continued to function much as it had in the past, in its own traditional cultural arenas. Mimetic skill still provides the cognitive basis for human social institutions like craft, athletics, dance, and the complex non-verbal expressive dimensions captured and cultivated in ritual, acting and theatre; and language provides the narrative framework that ultimately governs those institutions. Myth and narrative thought are the governing level of thought in oral cultures. Whether they know it or not, all humans grow up within a mythic system. Myths form the cultural glue that holds societies together. Myths and stories contain and supersede the prototypes and mimetic stereotypes of social roles, social structure and custom. They rely on allegory and metaphor, and lack precision, but they remain the universal form of human integrative thought, and one of the most potent and meaningful ways of representing reality.

In modern humans, language and mimetic skill work closely together in the expression of ideas, but can also be used independently of one another, to create simultaneously contrasting messages. Such contrasts are common devices in many areas of culture, but especially in cinema, theatre, comedy and opera, which employ mimetic–linguistic counterpoint very effectively The tension produced by driving these two contrasting modes of representation in opposite directions is a very powerful dramatic device. This suggests that these separate representational realms are sufficiently independent in the brain that they can operate concurrently, without interfering with one another.

THE THIRD COGNITIVE–CULTURAL TRANSITION: THEORETIC CULTURE

The third transition involved a switch from mythic to theoretic governance. The two evolutionary steps described above form the innate structural foundations of human thought, our gene-based cognitive inheritance. But cognitive evolution did not stop when we reached our modern form, somewhere between 100 000 and 50 000 years ago. A third major cognitive breakthrough has to be posited to account for the astonishing changes that have taken place more recently. These changes revolve around one central trend that has dominated the history of the past 20 000 years: the externalisation of memory.

Early humans, like their predecessors, depended on their natural or biological memory capacities. Thus, even though language and mimetic expression allowed humans to accumulate a considerable degree of knowledge shared in culture, the physical storage of that knowledge depended on the internal memory capacities of the individual members of a society. Thought was entirely inside the head; whatever was heard or seen had to be remembered and rehearsed orally, or visualised in imagination.

The advantages of external memory storage are obvious, but the invention of external memory devices has taken at least 20 000 years, and the full social realisation of the power of external symbols is very recent. The common keyword for the most recent phase of that transformation is literacy, but this term needs broadening to include much more than its conventional connotation, which in Western culture often means simply the ability to read and write alphabetic symbols. A more adequate description of human symbolic literacy would encompass all the skills needed to use every kind of permanent external symbol, from the pictograms and line drawings of the Upper Palaeolithic, to the astrolabes and alchemical diagrams of the medieval era, to the digital information codes used in modern electronic communications.

There has been no time for a genetic adaptation for external symbol use. We have basically the same brain we had 50 000 years ago. It might be argued that the shift to external memory was purely cultural, and therefore not as fundamental as the two previous ones. However, using the same criteria employed to evaluate earlier cognitive steps, recent changes constitute strong evidence for a third major breakthrough in our cognitive evolution. Both the physical medium and the functional architecture of human memory have changed, and new kinds of representations have become possible.

External symbols have transformed the medium of storage, although they constitute a change in technological, rather than in biological hardware. This is not trivial, since the storage properties of external media are very different from those inside the head. Whereas biological memory records are in an impermanent, fixed medium with a constrained format, external records are usually in enduring, refinable and reformattable media. These properties allow humans to construct completely new types of memory records, and to expand greatly the amount of knowledge stored in memory. External storage has also introduced new ways of retrieving and organising information; in fact, the retrieval workhorses of biological memory – similarity, and temporal and spatial contiguity – are not particularly important in external memory retrieval. The addition of so many external devices has actually changed our memory architecture – that is, the storage and processing options in the system, and their configuration – allowing us to move freely through an external information space that is virtually frozen in time. Because of their stable display properties, external memory devices have allowed us to harness the power of our perceptual systems, especially vision, for reflective thinking; and have literally changed which part of the brain we use to do much of our thinking. This has increased our options for interrelating various kinds of images and information, and for doing mental work in groups. All this has a neuropsychological dimension. There has been an invasion of the brain by culturally imposed programming, mostly in the form of institutionalised education.

A partial list of devices mastered by humans along the way to full symbolic literacy includes (in rough historical order) iconography, maps, emblems, totems, pictorial representations, pictographs, sequence-markers like knotted cords or prayer beads, various types of tokens, currencies, property markers, writing and counting systems, mathematical notations, schematic and geometric diagrams, lists, syllabaries and alphabets, scrolls, books, archival records of various sorts, military plans, organisational charts, environmental signs of various kinds, graphic images, scientific manuals, graphs, analogue instruments, specialised technical languages, computing languages, and a variety of modern multimedia storage devices that employ virtually all of the above. Even our personal memory system has been programmed with photographs, memoranda, TV images and other kinds of stored knowledge.

Once the required codes are in place in the brain, and the semantic memory system has a sufficient base of knowledge to work with, a successful external memory device will reproduce an intended mind-state in the reader or viewer. To the expert reader, the encoding strategies are so deeply established that the medium itself is invisible; ideas pop out of the page, and the message is processed unconsciously. While processing a major symbolic artefact – a novel, for instance – a particular set of abstract representations is set up in the reader's mind; and this temporary mind-state is highly dependent on the external device. Once the artefact is removed, little remains; put down a long novel and the temporary richness of the story rapidly dims, leaving only a general impression of the story and its characters. Pick it up again, and within minutes, the world created by the author reproduces itself in the mind.

The literate mind has thus become externally programmable, which is both an advantage and a danger. The advantage lies in the creative possibilities of symbols; societies can support much greater complexity, science and technology can advance, scholarship is made possible, and artists and writers become cognitive engineers, leading their audiences through tangled symbolically driven nets of ideas to end-states that are not otherwise conceivable. The danger is found in potential threats to individual integrity; free access to external memory tends to pull apart the unity of mind, fragmenting experience, undermining the simpler mythic thought structures humans have grown attached to, and exposing them to a bewildering variety of powerfully packaged messages.

The more complex forms of symbol use require the combination of all kinds of visual representations – pictorial, ideographic and phonetic – into large-scale external artefacts, such as architectural proposals, engineering plans, government commissions, scientific treatises, cinematic scripts or works of art. The high-level skills needed to do this kind of mental work are difficult to acquire, and far from being universal to all humans. These brave new capacities were not acquired without trade-offs. There is only so much brain-matter (or mind-matter). The physiology of brain plasticity suggests that when we increase the demands on one area of the brain it expands its territory more or less in proportion to the imposed load. Accordingly, cerebral capacity is used up and no longer available for something else. There is some evidence that with literacy we sacrificed a degree of visual imagination; and that we are losing our capacity for rote verbal skills, like mental arithmetic and memorisation (see for example Richardson, 1969). The nature of these trade-offs should be explored further, because symbolic literacy is not easily or naturally acquired in development, the way mimesis and speech are. Literacy is unnatural, and requires a wrenching redeployment of cerebral resources.

It is easy to underestimate the degree to which we depend on external symbols. Throughout a typical day, we encounter many of them. From clocks, to cereal boxes, to microwaves, maps, cartoons and road signs, our day is filled with digital, analogue and pictorial representations, as well as complex devices and artefacts, such as equations, poetry and computerised systems. The impact of external symbols on the brain, besides their ability to engineer our mind-states, is increased cerebral baggage in some domains, and a decreased load in others. Exograms (as opposed to engrams or internal brain representations) give us a permanent external memory record, and allow us to distribute cognitive work across many individuals. Moreover, their capacity for iterative refinement is unlimited, their retrieval paths are unconstrained, and perceptual access to them is very good. This gives them an advantage over engrams, which are hard to refine, have very few retrieval paths, and allow only very poor perceptual access.

CONCLUSIONS

A final point about the mediating role of consciousness. In our traditional theories, we have often defined consciousness as a rather narrow band of short-term memory storage, a window only a few seconds wide, within which we pass through the stream of experience that constitutes a lifetime. Long-term memory may contain everything we know, but it is unconscious, and thus useless to us as conscious beings, unless we can retrieve it into awareness. But between these two systems, there is a level of conscious processing that I have labelled 'intermediate-term' governance. This is a much wider, slower-moving form of working memory, which contains all simultaneously ongoing mental activity, including activity that is not as vividly conscious as, say, visual sensation, but is nevertheless a very active and causal element in behaviour. An example of this would be the complex of forces that govern a conversation between several people. Such a conversation has a vividly conscious element that is very short-term in its duration (the sounds of words as they are spoken). But it also has a slower-moving dynamic that can last for hours, and involves the strategic tracking of several lines of interconnected thought and the subtle adjustments made as the conversation itself unfolds in the memory systems of the participants. I have suggested that the existence of this slow-moving conscious process suggests that there is a dimension to brain activity that we still do not understand, and which must become the focus of a new generation of neurophysiological experiments aimed at slow-moving integrative neural processes.

The other focus of these studies must be the brain's interaction with culture itself. This will open up a whole new series of fields and specialties, but among these, the study of the neural effects of literacy training will stand out. When we introduce external symbolic memory storage into the traditional architecture of human memory, there is a radical change in the nature of the cognitive models we can propose: in effect, we reflect the entire internal structure of memory onto the outside world, and the conscious minds becomes a mediator between two parallel memory systems, one inside the brain, and the other (much larger and more flexible) outside. Of course, we retain our traditional biological memory structures, but we also have acquired a vast amount of permanent external symbolic storage, with novel retrieval and storage properties. Moreover, a new feature is introduced into the cognitive model: the external working memory field, which has become a very active area of research. External storage complements traditional biological working memory systems. For instance, if we sit in front of a computer, the screen becomes a temporary external working memory field. Anything we display in it is processed in consciousness, and the viewer is locked into an interactive loop with the display while creating, writing or thinking. This changes the traditional function of the brain's biological working memory system. In principle, this idea applies not only to computer displays, but also to other types of external symbolic display. For example, painters interact in this way with their easels, poets with their sheets of paper, accountants with their spreadsheets, and so on. The external memory device is built right into the cognitive system, and changes the properties of the system. The impact of such technology is even greater if the technology is an active player, as, for instance, in the case of an interactive computer display system with cognitive properties of its own.

This not only changes the way the growing brain adapts to the information environment, but it brings about an even more fundamental change in the existential predicament of individuals, as they accumulate massive networks of knowledge about the world, stored partly externally, partly internally. We might still be able to think of ourselves as 'monads' in the Leibnizian sense; that is, self-contained entities bounded by our skin membranes. But, as peripatetic minds plugged into a network, we are immersed in a gigantic external memory environment within which we can move around. We can connect with an almost infinite number of networks out there. We can share memory for a moment with other people and, at that time, we are networked with them. This creates new possibilities, some of which are not thrilling to contemplate, for group manipulation. But, it also provides more possibilities for freedom, and individuality, than at any other point in history. In simple oral cultures, freedom and individuality as we know them were virtually impossible. But in our complex cultures, there are so many ways of configuring the world that extreme individuality has become a possibility.

3 Consciousness and the limits of neurobiology[1]

HILARY ROSE

Consciousness has recently become an immensely fashionable theme within the new-found cultural popularity of the natural sciences. However, what is immediately noticeable about the proliferation over the past decade of books and journals with 'consciousness' in their titles or invoked in their texts is that they seem to be drawn to the cultural glamour of the concept, but with little sense that the concept of consciousness has an entirely other history. Consciousness seems to lie around in the culture like a sparkling jewel irresistible to the neurotheorists. There seems to be no recognition amongst the many biologists, artificial intelligencers, physicists and philosophers who have played in print with their new toy that the concept of consciousness is part of other discourses. Above all I want to underline that while for these neurotheorists consciousness is located within the individual human organism – and sometimes just the brain within that – the older tradition, coming from the humanities and social theory, sees it as located in subjectivity and intersubjectivity within a historical context. As David Lodge (2002) has pointed out, novelists may approach consciousness more readily than neuroscientists. The methodological individualism expressed in the objectivist language of the natural sciences erases both 'me' and 'you'; by contrast in social theory both agency and structure are crucial. For social theory there can be no development of individual consciousness without a social context.

[1] An earlier version of this paper appeared in Núñez, R. and Freeman, W. J. (eds.) (2000). *Reclaiming Cognition: The Primacy of Action, Intention and Emotion*. Devon: Imprint Academic.

The New Brain Sciences: Perils and Prospects, ed. D. Rees and S. Rose.
Published by Cambridge University Press.

My intention in this paper is to recall this other history, as I fear that much current theorising seems to have thrown the consciousness baby in all its subjectivity and sociality out with the ineffable bathwater. My hope is to foster a situation where we acknowledge these entirely different disciplines and epistemologies, and become more aware of the limits of our own discourses. Recognising limits may help friendly and productive conversations and offer the best possibilities for good theorising. For while the natural-science community has become fascinated by consciousness, social sciences and the humanities have become equally fascinated with both nature and the body. While it is evident that there are very different consciousnesses, natures and bodies in play, the new commonalities of concern can be read positively. Constructive conversations are possible.

FROM MARX AND FREUD TO FEMINISM, BLACK AND ENVIRONMENTAL CONSCIOUSNESS

The *Shorter Oxford English Dictionary* defines consciousness philosophically, as 'the state or faculty of being conscious, as a concomitant of all thought, feeling and volition' (earliest usage, 1678). As this fusion of reason, emotion and intentionality, consciousness was discussed only by those interested in the systematic exploration of subjectivity and intersubjectivity. Those in pursuit of positive objective knowledge of either nature or society necessarily either disregarded or dismissed the topic. Consciousness was the ineffable, perhaps to be graciously acknowledged as with qualia and the discussion of the redness of red, or Searle's ignorant translator of Chinese locked in a sealed room, but not as the subjective and intersubjective, to be seen as researchable by positivistic science. Thus for much of the last century the intellectual giants dominating the discussion of consciousness were Freud and Marx. However, their views of the gains to be achieved through enhanced consciousness could scarcely be more different. Freud's utopian vision was for individuals to exchange life-deforming neurosis for modest unhappiness. Marx's utopian vision was for nothing less than world revolution, to be attained by the collective and correct consciousness of the proletariat which would enable the workers to transform capitalism into socialism and thence communism.

As I sketch out below consciousness as a phenomenon only accessible through the exploration of subjectivity and intersubjectivity (that 'me' and 'you', where 'you' may be one or several), changeable, embodied, recognisably composed of feeling, cognition and intentionality, has long escaped the disciplined texts of Freud and Marx and their followers. The year 1989 dealt a fatal blow to the claims of actually existing socialism, whilst arguably selective serotonin reuptake inhibitors, whether in herbal or conventional medicine form, can modify depression, if not neurosis, more speedily than analysis. Nonetheless consciousness as subjectivity and intersubjectivity within a historical context, has both entered our everyday culture and been the focus of a tremendous amount of research in both the humanities and the social sciences not least through the memoir and autobiography (Stanley, 1992). So too have novelists, especially from the start of the twentieth century, attempted to catch the fleeting and evanescent nature of what William James called 'streams of consciousness', notably in the work of James Joyce and Virginia Woolf. Lodge, recognising that Gerald Edelman is one of the most sophisticated of the neurotheorists, puts his thesis into the mouth of his fictive cognitive scientist Ralph Messenger: 'That's the problem of consciousness in a nutshell . . . How to give an objective, third-person account of a subjective, first-person phenomenon.' His novelist heroine Helen answers for the whole literary tradition: 'Oh, but novelists have been doing that for the last two hundred years.' (Lodge, 2002: 29).

That nineteenth-century project of the young Marx was specifically to overthrow false theories of consciousness, above all Hegel's 'idealism' and Feuerbach's 'materialism', and to discover what he spoke of as the 'true role of consciousness' in history. Consciousness for Marx did not lie outside history but was immanent within it; thus for historical materialism, 'consciousness did not determine being but being determined consciousness'. It followed that the task of philosophers like himself was simply to explain to the world 'what its own struggles are about'. Above all this meant helping the proletariat advance from its necessary stance of collective self-defence (trade unionism) which he spoke of as limited to 'class in itself', to that conscious unity of revolutionary theory and practice of 'class for itself'. The powerful elaboration of *History and Class Consciousness* by George Lukács (1923; 1971) still makes astonishing reading eighty years on, not least in his reflections on the relations between political economy and violence.

From the present point, looking back over the last tumultuous and often savage century, few are left able to share in the Marxist proposition that the longed-for 'leap to the realm of freedom' depends on the capitalist world dividing into two great classes, with one of these, the proletariat, carrying the hopes of all humanity. It was not just that the theory failed to grapple with the harsh grip of nationalism, so that even while the Russian revolution was taking place the European working class was letting itself be slaughtered by the millions in a nationalistic and imperialist war, but that even despite Mao's ingenuity in recruiting the peasants as the necessary allies of, or substitutes for, the proletariat, somehow still that universal symbol of humanity, the proletariat in the old capitalist countries, remained deeply gendered and raced.

But what is central within both these struggles and the attempts to theorise them is the role of consciousness, as always a dynamic, never a static concept. Chinese peasants, speaking bitterness, debated 'Does the landlord need the peasant or does the peasant need the landlord?' in order to break through to a new level of consciousness. The capacity to overthrow the landlords required that the peasants became conscious of them as not merely exploitative but unnecessary. Shared discussion was the key to the transformation of shared consciousness. Franz Fanon's *The Wretched of the Earth* (1967) drew an inescapable connection between the development of Black consciousness and the necessity of revolutionary violence – an analysis which powerfully influenced US African Americans, not least Malcolm X. Steve Biko's Black Consciousness movement in South Africa, although committed to non-violence, was seen by the apartheid state as challenging white hegemony and hence requiring his murder. Despite the insights of the nineteenth-century women novelists, not least Jane Austen and George Eliot, politically women could only begin to grasp the specificity of their oppression within gender relations through closed 'consciousness-raising groups' in which experience was shared and analysed in order to find new words, new concepts, for 'the problem that had no name'. All these transformative movements have required a shared cognitive and emotional understanding of exploitation and oppression and the shared will to change it.

Feminist consciousness-raising, like Black consciousness-raising, was carried out by the oppressed themselves. Initially the project of global sisterhood seemed to offer the realm of freedom with every bit as much promise as the proletariat, but the problem of just exactly 'who' this sisterhood was appeared rather quickly. Challenges not only from Black feminists but also from post-colonial, older, younger, lesbian, disabled women, etc., argued against a monolithic construction that the needs and interests of white, middle-class, young, Northern, able-bodied, heterosexual women were equated with those of all women everywhere. Although feminists have grappled with the complexities of difference, in the context of hypermodernity it has been an extraordinarily difficult process. Those dreams of a shared universal consciousness through the common 'species being' of women have faded. Alliances between social groups including women cannot be read off from gender; instead they have to be built.

The usual clarity of hindsight enables us to see that such social movements could not succeed in their own Enlightenment terms. The social and economic gains have been modest. Poverty is still all too present in the lives of women and their children, whether in the world's richest or poorest societies. But despite this the cultural gains over the last three decades have been huge. Of course there are unreconstructed social groups and individuals, but what is extraordinary, is the immense change in gender relations since the mid-century. Those notions of femininity and masculinity once seen as 'natural' in Euro-America (like the Flintstones typically for ever frozen in a 1950s sub-urban version of the Stone Age) are now widely seen as historically constructed. A changed shared consciousness, thought, feeling and intentionality work to expose and name phenomena flowing from such gendered concepts of human nature. New words come into existence, old words are given new social meaning: 'sexual harassment', 'male violence', 'rape', 'racism', 'child abuse' (phenomena that had hidden themselves in nature – and legally in the private domain) are pulled into consciousness, culture and the public courtroom. Putting such concepts into culture calls the perpetrators to account. Rape is no longer discounted as part of masculine nature, but located in the intentionality of particular men. Even in war, that historic legitimator, not all men rape (see Chapter 8, this volume). But our failures to prevent these ills as swiftly as many of us hoped has also made us understand that changing subjectivity, changing the consciousness of abuse, is no simple matter for either a perpetrator or even a victim.

Nor has the huge shift in, say, gender or race consciousness been plain cultural sailing, not least because of sociobiology's immense efforts in the 1970s and beyond to reassert the claim that biology is destiny. What changed cultural and political consciousness names as intentional crimes, for this renewed Social Darwinism are seen as simply the working through of the demands of selfish genes. In the 1990s sociobiology has been rebranded and reconstituted as evolutionary psychology, and individual genes have ceded in explanatory power to evolutionary imperatives fixed in a universal but gendered human nature laid down in the Pleistocene; but the consequences for assumed universal human nature are not dissimilar (Rose and Rose, 2000).

For most of this century Marx's other great insight, that the natural world also has a history and both shapes and is shaped by humanity, was lost, relegated to a subordinate cultural current. Projects of human liberation were for many years cut off from the need to conceptualise or care for the ecosystem of which human beings are an integral part. Thus until the past three decades social movements – classically the working-class movement, but also movements of national liberation – have excluded any serious analysis of either nature or the natural sciences as knowledges about nature. Nonetheless what is clear and encouraging today is the widespread concern, particularly among children, for nature. A new environmental consciousness, however often it gets pushed onto the back-burner by economic imperatives that mean that the urgent pushes out the important, is an integral if uneven presence within our present political and cultural consciousness.

Not only has green nature 'out there' entered our consciousness, but so have our bodies. Today social theory understands bodies not as some fixed entity but as embodied experience. While beauty and adornment of the human form were always important in the ancient world, mortification/disregard of the flesh being only one tradition, today's bodies are dieted, exercised, pierced, handed over to cosmetic surgeons, fretted over in an unprecedented way in order to achieve conformity to new aesthetic visions of lean muscularity. Nature and the body are back, both in everyday cultural life and also in the academic discourse of the humanities and the social sciences. No longer are the constructions of nature and the body seen as the automatic and exclusive responsibility of biomedicine. Other discourses have a great deal to say – and this is perhaps what is making conversations difficult.

NEUROSCIENCE'S INTERNAL PHRENOLOGY

While natural scientists not infrequently rage against the (mis)-appropriation of their concepts, by for example New Age philosophising as in 'the quantum body', or by the Lacanian feminist philosopher Irigary (Sokal and Bricmont, 1999), they seem to feel that their own (mis)appropriations are entirely justified. How else are we to understand the way in which after a decade of vast funding for the neurosciences generating thick empirical descriptions of brain function with so far fairly modest theorising, the past decade has designated 'consciousness' as the hot theoretical topic? The topic has certain instantly notable characteristics, in that it is not pursued by brilliant young scientists, whether doctoral students or even postdoctoral fellows as in, say, the DNA story, but by the already famous. Career-minded young neuroscientists have been heard to describe an interest in consciousness, even while giving papers at a meeting of that name, as a CLM – a career-limiting move.

The main contributors to the consciousness debates have rarely been drawn from within neurobiology itself but are established figures in other related areas: biology (Francis Crick), philosophy (Daniel Dennett; Patricia Churchland), linguistic psychology (Steven Pinker) and physics/mathematics (Roger Penrose). Does this rather peculiar demographic and professional profile of middle-aged and more consciousness theorists reflect what experimental natural scientists self-mockingly call the philosopause, that change of life when experimental fertility runs out? Anthropology and literature suggest that it is unwise to dismiss jokes as either trivial or treacherous. A joke is typically a rich source of dangerous comment on matters that might be socially difficult to discuss head-on. It was Lear's fool who stayed with him. Let me briefly explore some influential examples of the new science-based constructions of consciousness. First, the Cartesian inheritance of the dominant tradition in the natural sciences means that most, philosophers and scientists alike, equate consciousness with cognition (Dennett, Flanagan, the Churchlands, even Searle among the philosophers, Penrose and the artificial intelligencer Aleksander). Daniel Dennett (1991) provides an almost classically one-dimensional concept of consciousness; thus what cannot be subsumed under his personal area of expertise, namely 'cognition', is excluded from consideration. It has been left to Damasio (1994; 1999) to insist that 'Descartes' error' is precisely that of focussing on cognition at the expense of feeling when theorising brain function and conscious experience.

Numbers of the new objectivist theories of consciousness evade the philosophical challenge of subjectivity. Crucially, where is the 'I' of agency? Jerry Fodor (1998) in his review of Pinker's (1997) *How the Mind Works*, with its modules for this and that cognitive function, echoes this in his question, asking where is the 'I' that holds the modules together. The objectivist stance leads many natural scientists to speak of consciousness as the equivalent of not being unconscious (Stuart Hameroff, Susan Greenfield, Gerald Edelman). Crick (1994) makes a virtue of the narrow concept. Consciousness, he argues, may be a hard problem, but if we reduce it to awareness it becomes more tractable and we can then take a segment of conscious experience, for example visual perception, and study this as a model for the whole.

This approach assumes that human subjectivity is the equivalent of that of any other creature that is capable of being either awake, asleep or unconscious. Without assuming that there is nothing in common between the neurobiology of different species, the idea that there is a one-on-one consciousness between species borders on farcical. It suggests that my cat Hypatia and I (who get on rather well and are both experienced at being awake and asleep and in sundry drowsy states between) have the same consciousness. Thus Susan Greenfield's metaphor of consciousness as a dimmer switch, in her otherwise lucid guided tour of the brain (1997), describes very well both our varying levels of sleepiness and wakefulness and their neuronal correlates; but this entirely avoids saying anything about the subjective consciousness of either Hypatia or myself. For that matter, who switches the switch? Greenfield is not alone. Koch and Crick share a not dissimilar metaphor of the 'searchlight' of consciousness (Crick 1994). The problem of theorising consciousness is evaded by the simple-minded acid of objectivism.

Elsewhere the explanations demonstrate a bizarre reprise of nineteenth-century phrenology. The partially localised but also internally compensatory brain functions shown up in imaging studies of the brain, while not unfriendly to the localising endeavour of phrenology, emphasise dynamics. By contrast claims by the new internalist phrenology seem as fixed as the nineteenth-century phrenologists' categories of 'amativeness' or 'philoprogenitiveness'. This new phrenology is not limited to knobs and whorls on the surface of the skull, but goes deep within the organ. Thus Crick suggests, at least semi-seriously, that 'free will' can be located precisely in the anterior cingulate sulcus of the cerebral cortex. Yet this insistence that 'free will' can be located in a specific brain region reads as a curious echo of nineteenth-century phrenology with its materialist longing similarly to locate highly complex behaviours, feelings and dispositions. The cultural reprise taking place within the discourse of neurobiology finds itself once more able to map into the brain highly complex social phenomena; today such categories as 'homosexuality' or 'free will' have replaced the 'spirituality' or 'amativeness' of our Victorian ancestors. As for philoprogenitiveness, the sociobiologists (notably Richard Dawkins) are firmly convinced that it lies in brain processes switched on by our selfish genes. Such reductionist materialism simply eschews context, not least the social and cultural.

While Crick positions himself within that 'Cogito ergo sum' assertion of Cartesian identity he also wants to find space for agency, or volition – in his terms, 'free will' (see Chapters 1 and 5, this volume). This preoccupation with 'free will' stems from the self-imposed problems which arise from his and other molecular biologists' belief in the determinism of DNA as the vaunted 'master molecule'. A similar problem besets others of the biology as destiny school (notably sociobiologists and fundamentalist Darwinians), who, having come to certain biologically inevitabilist conclusions, find themselves faced with highly negative social futures for human beings. Unwilling to accept these gloomy scenarios, they invoke free will, typically in the closing pages of their books, to escape the iron cage they have themselves constructed. Pinker's assertion that if he didn't like what his genes programmed him to do, he would tell them 'to jump in the lake' (1997: 52) is a classic in this sky-hook approach to free will.

AMO ERGO SUM

From the perspectives of, say, the history of ideas or anthropology such propositions sound extremely strange, for they assume that concepts such as 'free will' or 'homosexuality' exist outside history and culture and can therefore have a physiological location in the individual's brain. To historically informed discourses, concepts such as 'free will' are integral to specific cultures at specific historical periods, and indeed are typically attributed only of particular kinds of people within those cultures. Only as we move towards the twenty-first century does it even begin to be possible to think of free will as something to be found equally among both genders, rich and poor, slave and freeborn, and all 'races'. Without such a historical sensitivity a theoretical biologist must argue either that those 'Others' had the specific brain area but that there was no expression of the attribute, or alternatively that evolutionarily the Others developed without that bit of brain.

Thus in the public space occupied by men of a certain social class, Descartes claimed reason for himself and others like him in the universal name of Man. It was left to Rousseau in his *Social Contract* to spell out the destiny of female Others, stressing the affinity of women for emotion and to the private. In his utopian educational project, women were supposed to claim (that is if Rousseau had not excluded the classics from Sophie's education) 'Amo ergo sum.' My hunch is that Sophie's claim could make a theoretically better starting point for a less biogendered approach to consciousness. 'I love, therefore I am' offers feeling with and for others, as the condition of identity and consciousness. The first-person perspective is placed on an ontological level with an ethical awareness of others.

RECOVERING FEELING

Despite this general drift towards reductionism and objectivism there is currently a modest counter-current within this natural-science-based discourse of consciousness, an attempt to put emotion back in. From the animal psychologist's Nick Humphrey's tender if confused set of essays called *Consciousness Regained* (1983) to the terrific title if rather schmaltzy contents of Daniel Goleman's *Emotional Intelligence* (1996) or Joseph LeDoux's *The Emotional Brain* (1996), Antonio Damasio's *The Feeling of What Happens* (1999) and Walter Freeman's *How Brains Make Up their Minds* (1999), there have been a number of attempts from within the natural sciences to put feeling – that which was seen as the business of Sophie and her sisters – back into thinking.

Regrettably even the best of these attempts are carried out with little direct reference to feminist scholarship. While few can escape the feminist challenge, whether in our most intimate lives or in the most public of spaces, it remains possible for masculinist academic discourse to exclude feminist theorising. Yet part of the immense intellectual energy of feminism has entailed a critical revisioning of reason – rationality itself. Such a project is surely an ally of the attempt to expand consciousness beyond cognition. I will not even begin to trace that research programme, but go straight to its conclusion – one incidentally shared by many theorists of the environmental movement – which sees the asocial construction of rationality embedded in modern Western science as indifferent to the needs of people and nature alike (Rose, 1994). Both movements and their theorists seek to re-vision the concept of rationality, to make rationality socially and environmentally responsible. Bringing in the social, bringing in an ethical concern with people and nature is of course to admit subjectivity, caring rationality, and the shared will to do things differently. Taking feeling seriously helps an interdisciplinary discussion of consciousness to richer and more responsible places.

4 Mind metaphors, neurosciences and ethics

REGINE KOLLEK

Unravelling the mysteries of the mind is perceived to be one of the greatest challenges of the twenty-first century. The development of new biomedical and computational technologies and their application in brain research, which was accelerated during the 1990s by the US 'Decade of the Brain' initiative, led to an exponential increase of methods and possibilities to examine, analyse and represent the central nervous system and its workings on many different levels. These developments foster the vision that one day it may be possible to explain psychological and cognitive phenomena as the causal effects of brain chemistry.

At least for some neuroscientists such as the Nobel prize winner Eric Kandel, it is the declared goal to explore all classical philosophical and psychological questions related to cognitive functioning with the methods and concepts of cellular biology. Other researchers think that one day it may even be possible to locate the origin of religious feelings and spirituality in the structures of the brain. According to sociobiologist Edward O. Wilson, the final goal of scientific naturalism will be reached if neurosciences succeed in explaining traditional religion as a material phenomenon. Wolf Singer, a leading neurophysiologist, claims that in the light of the new neuroscientific insights we have to say goodbye to our traditional understanding of human freedom. Even if only parts of such far-reaching claims are realistic, it can be anticipated that the advancement of modern neurosciences will not only change current beliefs about fundamental phenomena of the mind, but our conception of humans in general. Therefore, I agree

The New Brain Sciences: Perils and Prospects, ed. D. Rees and S. Rose.

with Steven Rose who said that 'one thing is certain: in the coming decade the brain sciences are going to be as much at the forefront of public social, ethical and even religious debate as genetics' (Rose, 2000: 98).

One of the key questions is what norms and policies will guide the future development of the neurosciences and the application of neuroscientific knowledge in medicine and society. Current discussions about the social and ethical implications of the new sciences and techniques of the brain focus around issues raised by the clinical neurosciences such as experimentation with patients who are unable to give informed consent, psychiatric or behavioural genetics, or – more recently – neural stem cell transplantation. Most of them concern the dignity of the person and respect for autonomy. But the increasing abilities to represent the brain and its workings and to interfere with its functions evoke issues that reach beyond the individual, affecting human self-perception and having relevance for society and humankind in general.

Defining norms is a matter of ethics. Traditionally, ethics as a philosophical discipline probes human action and critically examines whether it can be justified, whether it conforms to existing norms and standards and whether it is free of logical inconsistencies and internal contradictions. Today, ethics is increasingly occupied with questions posed by medical and scientific developments. When organ transplantation became a treatment option, society was forced to redefine death in a way that was adequately compatible with normative intuitions, ethical reasoning and the practice of transplantation medicine. In this process, the neurosciences played a central role. In contrast to everyday knowledge and perception of death, which relies on what everybody can see or feel, like cessation of heartbeat or a body becoming cold, a new definition – 'brain death' – was established, which was informed by the concepts, findings and proofs of clinical and theoretical neurosciences. Therefore, in the context of transplantation medicine, 'ultimately, our very definitions of life and death are dependent on the findings of neuroscience' (Blank, 1999: 43). They are granted authority to identify phenomena of the brain, which can be correlated to mind phenomena and serve as points of reference for normative decisions of society.

Normative statements about the legitimacy of specific interventions into the human body or brain therefore refer to neuroscientific knowledge and concepts – a phenomenon that can also be called the 'self-reference' of neuroethical reasoning. This reference to such scientifically defined concepts as 'brain death' raises important questions concerning their epistemic status. Since they do connect neuroscientific knowledge and ethical reasoning, their epistemic status needs to be thoroughly scrutinised. In this chapter, I shall highlight some of the problems pertinent to this self-reference of neuroethical reasoning. I begin by describing some of the features of neuroscientific concepts which are employed in the discourse on ethics, then turn to the epistemic status of such concepts, and finally draw some conclusions for the future debate on ethics and neurosciences.

THE HUMAN 'SELF': A POINT OF REFERENCE FOR ETHICAL REASONING?

In ethical discourse concerning the neurosciences, many philosophers and ethicists are critical about interventions or manipulations of the human brain because such interventions may alter the human self. The integrity of the self therefore serves as a normative criterion in order to distinguish legitimate from non-legitimate interventions into the human brain. The necessity for defining such criteria has already become an issue in the context of gene therapy designed to treat diseases of the central nervous system, or of transplantation of fetal neural cells into the nervous system of patients with Parkinson's or Huntington's disease. It will become even more important if and when embryonic stem cell transplantation becomes an option for the treatment of degenerative or other diseases of the human brain (see Chapters 12 and 13, this volume), or with respect to the growing capabilities of psychopharmacology (see Chapters 14 and 15, this volume).

Not to interfere with what is called the self of a person (i.e. the very perception the individual has of himself or herself), however, is a norm that refers to something difficult to grasp or to define. If the 'integrity of the self' should serve as a criterion in ethical reasoning related to interventions into the brain, we first have to describe or to define what we are referring to if we talk about a person's self.

Self-reflections: the self in the mirror

The self has been an object of study in different disciplines. Philosophy, psychology and sociology have produced voluminous descriptions and analyses of the subject (see also Chapter 3, this volume). During the last years, the neurosciences too have become increasingly interested in analysing processes of the brain thought to be related to aspects of a person's self. The neuroscientific study of the self, however, is just beginning to emerge. Several neuroscientific disciplines have tried to identify parts or regions of the brain that may be involved in the development or expression of processes related to the self, also called 'self-processes'. One approach uses functional magnetic resonance imaging (fMRI), which detects regions of the brain showing increased blood flow as a result of changed circumstances. This is supposed to be an indicator for neural activity (see Chapter 10, this volume). This is therefore an important method of choice, if one wishes to observe the workings of the brain directly. Although their techniques and their interpretations are in their infancy, some neuroscientists have already begun to probe the neurobiological underpinnings of complex phenomena related to the self.

Defining the concept of self and understanding the cortical underpinnings of such a concept is a huge challenge. As a first approximation one can say that it consists of different entities and processes. One phenomenon, which is supposed to be involved in the development of the self, is self-recognition of one's own face. The face is our most characteristic and most personal external feature. Recognising it seems to be unique to primates closely related to humans, and is limited to humans older than 18 months and adult great apes like chimpanzees or orang-utans. It has been shown, for instance, that processing of self-relevant information differs from processing objective information, as distinct cerebral areas are activated. In recently published findings from experiments with subjects watching images of themselves, scientists could identify by fMRI increased blood oxygen levels in specific brain areas. They suggested 'that a neural network involving the right hemisphere in conjunction with left-sided associative and executive regions underlies the process of visual self-recognition' (Kircher *et al.*, 2001). Similar, although not always identical, results were reported by others.

It is not completely unexpected that a complex process like self-face recognition involves different regions of the brain, although the prefrontal cortex may play an important role in such processes. But self-perception certainly comprises many more aspects than self-face recognition including self-evaluation and autobiographical memory. Furthermore, there is no explicit consensus yet on what qualities belong to the self, and how to conceptualise it. For Stephen M. Kosslyn, a cognitive neuroscientist from Harvard University, 'the self can be understood to arise from an interaction of information in memory, processing abilities, and temperament' if one assumes that mental events correspond to brain function (Kosslyn, 1992: 38). Furthermore, the individual builds up a self-concept, which corresponds to a special kind of information stored in memory. It functions as a kind of 'internal model' that can be used to guide behaviour (Kosslyn, 1992: 40).

Such an understanding of the self is based at least on two presumptions.

- The first is that mental events related to self-perception correspond to specific brain processes. Hence, they rely on what is called the 'doctrine of psychophysical parallelism', which underlies all modern brain research. This is the idea that the brain and the mind – and therefore mind-related processes – operate in separate spheres but still somehow in tandem. Though this doctrine is plausible and held to be

'true' by most neuroscientists today, it is not yet proven in the scientific sense of the term, since many of its conceptual and epistemological problems are unresolved. Therefore, although interventions into the genetic, physiological or anatomical structure of the human brain can interfere with its functioning, it is not at all clear in what way they can or do interfere with the self of a person. Or, in other words, it is unclear what are the causal relationships between the material structure of the brain and its constituents on the one hand, and such mind phenomena as the self on the other hand.
- Second, in the model of Kosslyn, the self (or self-concept) of a person not only depends on the interactions of different brain functions (memory, processing ability), which are themselves not very well understood, but also on 'temperament'. Temperament, however, is again composed of different physical, intellectual, emotional and moral qualities, which make up a person's personality, his or her attitudes, and behavioural responses to varying life situations. It can be understood as a mixture of motivational drives, which are influenced by, among other things, the endocrine and vegetative system. Therefore, temperament may not be a function of the brain at all, but the result of other physiological variables such as hormones, age, health, etc.

Even if the self were constituted by these functions only, it would be extremely difficult or almost impossible to assess which mechanical, chemical or genetic intervention into the human brain (or even into the human body) would interfere with or alter the self of the human person.

The central question for a debate on which norms should guide modern neurosciences, therefore, is whether mind phenomena like the self can serve as a critical point of reference to distinguish ethically legitimate from illegitimate brain interventions. Not only because of the complexity and flexibility of what is believed to constitute the self, but mainly because of the conceptual problems, this seems to be rather doubtful. If the self is conceptualised as an emergent feature of the brain and its neuronal architecture then it would probably be possible to pinpoint this feature by analysing and describing structural aspects of the brain and its cellular composition and biochemical workings. At least in theory, such a concept would make it possible to predict the effects of brain interventions on the self. However, if the self is conceptualised as resulting from interactions between the workings of the brain and external stimuli, then it would be rather difficult to forecast the effect of such brain interventions.

In the eyes of others: the social self

Self-evaluation also counts as a self-related process. In general, the term self-evaluation signifies processes by which a person addresses himself or herself in order to review or to assess certain personal aspects, be it behaviour, health, or some other characteristics of the body or the mind. Such an assessment, however, is dependent on (internal or external) points of reference, against which the rating or evaluation must be judged.

Evaluation of one's own social behaviour could be seen as an element of this process. Anderson *et al.* (1999) describe two patients who experienced injuries in the prefrontal cortex early in life. The injury led to what can be called a loss of moral sensibility, as well as of the ability to feel guilty. Despite normal basic cognitive abilities, neither of these people was able to realise rules of social behaviour. It seems that a defect in brain anatomy leads to a kind of 'moral blindness'. They could not retrieve factual complex and socially relevant knowledge. They showed defective social and moral reasoning, suggesting that the acquisition of complex social conventions and moral rules had been impaired. Apparently, neither suffered from their inadequate social behaviour; they did not even realise its inadequacy, since they did not have – or have access to – a reference point. 'When similar injury was suffered by people in adult rather than early life, however, it did not result in such far-reaching social impairments.' Anderson *et al.* (1999) therefore suggested that the lack of a sense of remorse or guilt relative to conduct indicates that early-onset patients were never able to acquire the emotional knowledge presumed to bias the reasoning process, thus options for actions cannot be qualified in emotional terms, and the consequences of future choices cannot be evaluated.

For many neurologists, this indicates that a functioning prefrontal cortex is essential for the acquisition of social and moral knowledge. If its functioning – and the linkage between perception and emotion – is destroyed, than learning, which requires emotional corrections, cannot take place. In his book *The Feeling of What Happens*, Antonio Damasio (1999) seeks to delineate the nature of consciousness and the biological source of our sense of self. His goal is to understand how we cross the 'threshold that separates being from knowing'; that is, how we not only know things about the world, via our senses, but how we are aware simultaneously of a self that is experiencing this 'feeling of what happens'. He postulates the crucial roles that emotion, memory and 'wordless storytelling' play in our conscious existence.

These findings provoke far-reaching questions: what about individuals who have no injury, but some sort of (genetically caused) physiological malfunction in the prefrontal cortex? Should one expect weakly developed moral behaviour, by analogy to persons being shortsighted due to a functional problem of the cornea? Nobody would blame blind people for their blindness – could we blame people with physiological malfunctions in the brain for morally aberrant behaviour? Or – the other way around – would morally aberrant behaviour have to be reinterpreted as being 'caused' by structural features of the brain and/or its neuronal constituents? If results and theories deduced from neurological findings are confirmed, we certainly would have to re-evaluate not only psychopathological cases of patients locked up in psychiatric institutions, but also the social challenges of neurosciences in general (see Chapters 5 and 6, this volume). Although there may be no genes for moral abilities or sensibilities, there may be structures or functions in the brain associated with specific processing abilities linked to moral behaviour. The ability to learn and recognise moral rules and systems seems to have a physical correlate: our moral sensibility – this highly respected achievement of human culture – seems to depend directly on the functioning of specific nerve cells or brain regions. Anderson *et al.*'s patients were not aware of their harmful behaviour, because they could not link responses from their social environment to internal feelings. In that sense, it can be said that this specific aspect of humanness or the human 'self' – the ability for moral behaviour – is localised in specific anatomical structures of the brain.

But this is only one perspective from which to look at the phenomenon of moral behaviour. If we stick with it, than we may be convinced that it is located in the brain and nowhere else. But what these findings also show is that the ability to make moral judgements and to differentiate between good and evil in people without lesions in the prefrontal cortex depends as much on the structures of the social environment and its responses to the actions of the child, as on the functioning of specific brain structures. If there were no adequate reactions of the social environment, the child – though capable – would not learn the difference between socially accepted and disapproved reactions and behaviours.

From this perspective, Damasio's findings correspond both with recent philosophical perceptions on moral learning and to empirically based sociological theories of the self, perceiving it as constructed in social interactions and undergoing biographical changes. If we see it like this, the self is neither a result of abstract (philosophical) generalisations, nor something located exclusively in the brain, but a result of interactive social processes that are memorised and constantly evaluated. They shape our perception of what we are, were we came from, how we think about ourselves and our families, how we act, and how we are perceived by others. Therefore, human self-conception cannot be described in terms of anatomical brain structures or qualities unchangeably attached to the individual only, but must be understood as a result of a process which depends as much on the presence of intact brain structures as on the perception of other members of society. Or to put it differently: the self and self-perception have to be described in a language of (changing) relationships, and not of attributes only.

The conclusion that can be drawn from these reflections is that empirical findings from clinical cases or experimental studies may be able to grasp some of the multiple dimensions of complex phenomena of the mind, but they exclude important others that cannot be described in the language of experimental neurosciences. Therefore, in order to make well founded assumptions about the nature of the self, the judgement has to be instructed by other disciplines. This is in contrast to the reductionist programme in the cognitive neurosciences, which aims at describing in the terms of cellular neurobiology all classical philosophical and psychological questions related to mental functions (see Chapter 1, this volume). Damasio's findings may be relevant for certain pathological cases of morally aberrant behaviour, just as the alterations in the huntingtin gene are relevant for the development of Huntington's disease, and they may also offer clues as to what neurobiology can contribute to an understanding of social and moral knowledge acquisition. They may not help much, however, in understanding specific social behaviours such as xenophobia and the social violence it motivates. The problem is that norms, which come from the realm of the social – the definition of deviance, good or evil – are projected into the material structures of the brain without knowing the means or the mechanism mediating this projection.

This leads back to the question of whether the integrity of the self is an appropriate criterion for ethical reasoning. The self is a very flexible, fuzzy concept, which is not at all clearly defined, either in the experimental neurosciences or elsewhere, although different disciplines have made valuable contributions to understanding what may be meant by the term. It encompasses different processes and aspects, which can hardly be isolated from each other and are difficult to examine separately. It remains true, though, that the concept of self is attractive, bridging as it does different realms of perception and experience.

BRIDGING UNLIKE THINGS BY METAPHORS

Psychophysical parallelism, which is the foundational doctrine of modern neurosciences, assumes that for every mental state there is a correlative nervous state. As sociologist Susan Leigh Star pointed out, one of the major difficulties of this doctrine is how to bring the two realms – the very concrete, fleshly brain, and the very abstract, invisible mind – together (Star, 1992). Due to the fact that the abstract object (the self, thoughts, feelings, memories, etc.) is inaccessible to physical examination and even the concrete object – the brain – is difficult to observe, many more problems are posed than one can even summarise in a short paper. Pragmatic research strategies developed a shortcut: instead of observing the brain directly, they look at substitutes that are thought to reflect changes in the brain: behaviours, movements, reports of emotions or thoughts, and cognitive performance. Imaging techniques now enable us to monitor physiological activities and changes in the brain more directly. What we observe, however, are not cognitive processes or the mind, but electric signals or patterns of blood oxygen and flow, which are, or may be, correlated with mind activities. They may at best show us how the processes of self-recognition or other mental goings on are organised superficially. We still have no idea of how they may be related to the 'qualia' – the subjective part of mind activities, although such feelings and experiences may lie at the heart of what we usually mean by the 'self', regardless of how well such experience reflects underlying brain events.

Since processes associated with the self and other phenomena of the mind cannot be measured directly, the terms and concepts used to describe them are empirically underdetermined. Furthermore, concepts designating specifically human features that are perceived as higher brain functions, are hybrids of unlike things: of something very abstract and something very concrete, of individual and social aspects of a person, of internal feelings and external perceptions. There is still no consistent scientific concept available that can adequately link phenomena from those categories together – only different modes of description, coming from various disciplines, perspectives and experiences. The neuroscientific conceptualisation of the self – perceived to be composed of several components resulting from neurochemical reactions – therefore is not a scientific concept derived from modelling of empirically observable phenomena, but serves as a proxy for something that cannot be described adequately in the language of neurophysiology and neurochemistry.

Such descriptions or analogies used to generate models of the self are nevertheless helpful. They are heuristic tools for the investigation of mind processes, and they are a way of building a bridge between unlike things without being able to describe their relation in analytic or systematic terms. Hence, they are metaphors. This sounds strange, but it is not. Scientific thinking and conceptualisation has much to do with building models, involving abstraction, idealisation and generalisation. It has to do with using analogies or metaphors in order to gain an idea of the internal workings of the brain and its capacity to generate qualia. It was this way that Robert Hooke, one of the first pioneers of modern science in the seventeenth century, approached memory and tried to find a scientific explanation of this phenomenon and its workings. In a famous lecture he gave before the Royal Society in 1682 he discussed a mechanistic theory of memory. The text, which was published under the title *An Hypothetical Explanation of Memory: How the Organs Made Use of by the Mind in its Operation may be Mechanically Understood* is full of metaphors and analogies, mostly derived from physics and chemistry. For example, for him Bolognese stone served as a metaphor for the brain, the physiological substrate of memory. Bolognese stone is a heavy spar. The combustion products of this mineral store light at daytime and emit it in the dark. According to Hooke, the brain has analogous properties: it stores memories like Bolognese stone and emits them at a later time (Draaisma, 2000).

Today, neurochemical and neurophysiological processes and the terms designating them have taken over the role of ancient metaphors. Like Bolognese stone, which was used as a metaphor for memory, but which represented only one feature of this specific brain phenomenon – the storage and reactivation of memory – neuroscientific terms and models also refer to one dimension of the self only: the workings of the brain. They seem to generate the self in the same way that the gall-bladder generates bile, or pancreatic islet cells produce insulin. Such models or comparisons are insufficient, as many neuroscientists acknowledge. They do not explain how the different dimensions of the self are connected or integrated: the material, the psychological and the social.

When models representing complex or multidimensional phenomena, or metaphors used to describe them, are communicated to other scientists or scholars and popularised, they often become transformed, simplified and reduced. They structure perception and cognition, and convey meanings. They also become aesthetically conventionalised, and value-laden. This helps shape scientific knowledge so as to make it accessible also to non-specialists. In that sense, such models also serve as boundary objects. In their meanings, they are adapted to the needs of philosophers as well as to the ones of neuroscientists studying brain processes occurring in parallel to the inner feelings of the subject. Although they have different connotations in both worlds, they are nevertheless identifiable by both communities and they are used as means of translation between different disciplinary languages.

Such models or metaphorical constructs therefore not only mediate between the material and the immaterial, but also between the factual (objective) and the normative, value-laden dimensions of the world. Although science aims to be as precise and objective as possible, the construction and use of models and metaphors seems to be inevitable. Therefore, trying to avoid them may be futile. However, metaphors can also bring with them potentially dangerous conceptional baggage, which may undermine their utility. The point is, that we have to use them critically, because their consequences can be very far-reaching, defining the limits and boundaries of a way of looking at the world, at human life and human bodies and intervening with it.

NEUROSCIENCES, METAPHORS AND ETHICS

At this point it becomes clearer how our image of the brain – its workings and what emanates from it – is linked to ethical questions. Philosophers and ethicists often refer to concepts like the human self when they discuss limits to interventions into the human brain. In that case, neuroscientific knowledge claiming priority in defining such concepts becomes central for judgement. As we have seen, neuroscientific concepts of the self have, like others, specific limitations in that they include and describe as relevant only what can be detected by the methods employed. Hence, other aspects and dimensions constitutive for the self but not detectable by these methods, may be excluded from conceptualisation. But if human judgements, which should be the basis for ethical assessment, are relying exclusively on scientific concepts about the structure and functions of the brain, then this results not only in reductionist but also circular reasoning. It is reductionist, insofar as psychological and social dimensions of the self are not included, and it is circular, as any answer to the question of which interventions into the brain may be permissible ultimately refers to neuroscientific concepts defining self-relevant processes and therefore predetermining the type of intervention that may or may not interfere with the integrity of the self. If ethical judgement relies on a perception of self-processes that are defined by the neurosciences only, it may be blind with respect to the limitations of such concepts and the values built into them and thus reproduces the uncertainties inherent in scientific knowledge, which is always 'in the making'.

Of course, this problem of normative discourse referring to current scientific findings and concepts is not unique to the neurosciences but is found in other fields of ethical reasoning as well. With respect to the brain, however, it becomes especially relevant, since the brain is seen as an organ essential not only for our physical integrity but also for our self and our identity as a person and as a human being.

In order to solve this problem posed by the flexibility of knowledge to which ethical reasoning refers, it has been suggested that ethics has to change with the progress of science. However, in this case ethics would become a derivative of science, as in the claims for 'evolutionary ethics' and consilience made by Edward O. Wilson and others. But this contradicts its very nature as a critical theory of morals. Confronted by the problems posed by referring to metaphorically loaded neuroscientific concepts – can ethics have a critical role in directing neuroscientific research?

With respect to human genetics and artificial reproductive technologies bioethics has a major role in guiding biomedical research. The demand for advice in ethics is increasing constantly. It is not so much articulated by moralists, cultural pessimists or critics of modern biomedicine as by the protagonists and promoters of the scientific and technological developments themselves. On the one hand, the establishment of bioethics originates in this growing demand for procedures and criteria for rational decision-making in the context of modern medicine. On the other hand, it is (in the USA and elsewhere) a reaction to social conflicts on biomedical issues. Therefore the advent of modern bioethics is directly connected to controversial innovations in medical technology and biomedicine. This is especially true for innovations that may collide with traditional norms and grown structures of human self-conception.

In that context, bioethics seems to become the authority guiding the technically feasible through the difficult waters of traditional philosophy on the one hand, and folk morals, desires and emotions on the other. Biotechnologists, geneticists and neuroscientists have an interest in controlling those social mechanisms intended to provide legitimisation for their work. This requires that moral limits have to be adapted to the needs of biotechnical innovation. By semantic redefinitions, which are facilitated by referring to poorly defined concepts, the realm of what is protected from manipulation or commercialisation often becomes restricted to what is not yet technically feasible or commercially exploitable (as in the debate over human cloning or 'designer babies'). Such reconstruction of ethical norms contributes to closure of social conflicts, and opens up new territory for exploitation, which would not have been available without this management of acceptance by bioethics (Feuerstein and Kollek, 1999).

With respect to the regulation of contested bioscientific and biomedical innovations bioethics has shown itself to be highly flexible and enormously successful. Although it does not guarantee the new technosciences of the body and the brain to progress smoothly, it increases the chance that the margins of what is accepted as morally legitimate are widened. It is this capacity for flexible adaptation to new developments and problems – connected to sensibility towards claims or demands guided by special interests, and for cultural sensitivities – that renders modern bioethics especially suitable to guide social adaptation to new scientific developments.

Therefore, I am not very optimistic about the capacity of bioethics to offer clear guidance on those implications of modern neurosciences, which lie beyond simple clinical interventions. Whereas the latter can in principle be regulated by the criteria of medical ethics (e.g. autonomy, beneficence, justice, etc.), the former escape from such strategies, which aim at the individual, since they pertain to groups of people or society. The neurosciences unfold – like genetics and the artificial reproductive technologies – a life-structuring reality which textures or even governs how we conceive of ourselves, and how we act. They not only aim at new or revised descriptions of human beings, but also at defining anew the basis of social relations and interactions in the terms of neuroanatomy and behavioural sciences. What is at stake is not only at the core of the subject, but also at the core of the social.

This reveals the ultimate significance of the neurosciences. They analyse and conceptualise the human brain, which is perceived as the physical locus of what it means to be human. Furthermore, if the brain is the locus at which the biological basis of our ability for transcendence and for symbolisation is localised, then it may also be identified as the organ wherein the origins of society may be found. Any description in neuroscientific terms of what it means to be human therefore affects society as a whole. As we have seen, the study of the brain and the knowledge derived from it is intimately connected to the preconceptions, goals and values of the neurosciences. Society as a whole, or parts of it, may not share the same preconceptions, and its goals may not coincide with those of science. As the neurosciences advance it is ever more vital for the future interests of society that we succeed in reconciling the perceptions of what it means to be human from a neuroscientific point of view with those from everyday life.

5 Genetic and generic determinism: a new threat to free will?

PETER LIPTON

We are discovering more and more about human genotypes and about the connections between genotype and behaviour. Do these advances in genetic information threaten our free will? This chapter offers a philosopher's perspective on the question.

Whether or not genetic discoveries do really threaten free will, many feel threatened, and it is not difficult to see why. If genetic advances enable us to predict with increasing accuracy and reliability what people will do, this seems to undermine the pretensions of individual autonomy and agency. In what sense do I choose for myself what I do, if you can say reliably in advance what that choice will be?

The free will dilemma is a hardy philosophical perennial. After thousands of years of work there is still no generally accepted solution, no clear demonstration that free will really is possible. A philosopher may well wonder how new genetic knowledge could make things any worse, or indeed make things any different.

THE SCEPTICAL DILEMMA AND DIMINISHED RESPONSIBILITY

To see why a philosopher might suspect that genetic information could not possibly make the problem of free will any worse than it already is, we need to consider the classic free will dilemma, an argument with three very plausible premises and a depressing conclusion.

The New Brain Sciences: Perils and Prospects, ed. D. Rees and S. Rose.

First, everything that happens in the world is either determined or not. Second, if everything is determined, there is no free will. For then every action would be fixed by earlier events, indeed events that took place before the actor was born. Third, if on the other hand not everything is determined, then there is no free will either. For in this case any given action is either determined, which is no good, or undetermined. But if what you do is undetermined then you are not controlling it, so it is not an exercise of free will. Finally, we have the conclusion: there is no free will. The argument has the form: heads or tails, if heads you lose, if tails you lose, therefore you just lose. Either determinism holds or it doesn't, if determinism holds there is no free will, if it does not hold there is not free will, therefore there just is no free will.

The dilemma is remarkably simple, and it packs an immediate punch. Let me nevertheless add a few comments on its structure and elements. The argument is clearly valid, in virtue of its form. To say that an argument is valid is not to say that its conclusion is true, but just that *if* the premises are all true, then the conclusion must be true as well, or equivalently that it is impossible for all the premises to be true yet the conclusion false. So anyone who wishes to reject the conclusion must also reject at least one of the premises. The argument does not assume any particular facts about our world, which suggests that the problem lies not in our world but in our concept. If the free will dilemma is sound – that is valid and with true premises – it seems to show that the very notion of free will is incoherent, something that could not possibly exist, a round square.

The first premise is indisputable, since it has the tautologous form P or not-P – everything is determined or not everything is determined. (Note that this is not the same as the disputable claim that either everything is determined or nothing is.) Just what determinism entails is a much more difficult question, and there are several different versions of the concept that could be deployed, though the first premise remains a tautology whichever one is chosen. The two most common versions of determinism appeal to causation or to the laws of nature. Thus determinism may be taken to be the view that everything that happens has a cause, or the view that everything that happens follows necessarily from the laws of nature in conjunction with the full state of the universe at any single moment. In fact this yields more than two conceptions of determinism, since the concepts of cause and law have themselves been given diverse philosophical treatment. Thus, some suppose that a cause is a condition sufficient for its effect, while others claim rather that it is necessary, something without which the effect would not have occurred. And while some philosophers have supposed that laws are simple de facto regularities, others have claimed that laws describe what happens by necessity, what could not have been otherwise.

The second premise of the dilemma, which asserts the incompatibility of free will and determinism, lacks the iron-clad security of a tautology, but there are powerful considerations in its favour. Free will seems to entail that the actor 'could have done otherwise', while determinism rules this out. The incompatibility of determinism with 'could have done otherwise' is particularly clear when determinism is defined in terms of the laws of nature (van Inwagen, 1975). If determinism is true, then what I did is entailed by laws of nature along with some particular facts about the state of the world before I was born. To have the power to have done otherwise, I would have to have the power either to change the laws or to change those prenatal facts. Clearly neither is possible.

Those who have tried to show that determinism and free will are nevertheless compatible have typically observed that the claim that my action was determined is compatible with my desires being among its causes and so that I would have acted differently, had my desires been different (Ayer, 1954). But defenders of the second premise reply that this is not enough to show that I could have done otherwise, if my desires are themselves just intermediate links in a long deterministic chain stretching back before my birth. In such a case, that people would have acted differently had their desires been different seems no more to show that they could have done otherwise than would saying that they would have acted differently, had the weather been different. Neither circumstance shows they have the power to change what they would do.

Another way of resisting the second premise is to question the connection between free will and 'could have done otherwise'. The desire being a cause of the action – which determinism allows – is clearly insufficient for free will. The addict is a model of someone whose free will has been compromised, though the addict desires the drug and that desire affects behaviour. But it has been suggested that what rules out free will in such cases is not that everything is determined, or that the agent could not have done otherwise, but rather that the addict does not have desires that are related to each other in the right way. For example, it has been claimed that the addict lacks free will because his desire for the drug is determined by the drug itself, rather than by higher-order commitment to wanting the drug (Frankfurt, 1971). Even if the addict is strangely happy to crave the drug, the craving is caused by the drug, not by the desire to crave. Ultimately, we all have desires we do not choose, but on this view what enables those of us who are not addicts to enjoy free will is that many of our desires are maintained because they are themselves desired. The defender of the second premise will not be satisfied by this, however, and will insist that the harmony of our mental economy is not enough to make room for the possibility of free will, if that entire economy and the actions it generates were determined by things that occurred before we were born.

The third premise of the dilemma is that free will is not compatible with indeterminism either. Supposing that some of my actions or their causes are themselves uncaused or ungoverned by deterministic law may allow that my actions could have been otherwise, but it does not seem to allow that *I* could have done otherwise. Indeterminism does not allow the agent to control her actions in the way free will requires. I do not exercise free will if my arm spontaneously rises, nor is the situation any more promising if we construe an indeterministic process as one that is irreducibly probabilistic, rather than one that is entirely random. We do not create room for free will by leaving desires undetermined or by loosening the link between desire and action.

The simplest explanation for the conspicuous absence of philosophical progress on the problem of free will is that the sceptical dilemma is sound: free will really is impossible. If that is so, then the answer to our questions about genetic information is simple, if pathological. Nothing can threaten what could not exist anyway. If free will is impossible full stop, then it is something genetic knowledge can neither reduce nor destroy.

But we may be unable to accept the sceptical dilemma, even if we cannot see exactly what is wrong with it. As Isaac Bashevis Singer is reported to have said, 'Of course I believe in free will. I can't help it.' Our disposition to treat others as free agents seems impervious to argument. The dilemma may show that our full-blooded conception of free will is incoherent, and that we must pare it down if we are to believe in something that might exist. The big question is whether this process would leave us with something still strong enough to support the use we make of the concept, and the connections we make between judgements of freedom and judgements of responsibility and dignity.

Here as elsewhere in philosophy, I think that we ought to be opportunists, willing to vary our standards to suit our purposes. Free will is not the only area where powerful reasons are given for incredible conclusions. In the theory of knowledge, for example, all the best arguments seem to show that we have no justification for what we are quite sure we do know, that the sun will rise tomorrow or indeed that anything exists outwith our minds. Taking those arguments seriously helps us to illuminate our cognitive practices, but it is also important to vary the setting on the 'sceptic dial'. Supposing the worst – that we can know almost nothing – is a way of revealing some strata of our belief practices, but for other philosophical purposes we must take some layers of our knowledge for granted.

Similarly, while for some philosophical purposes we may wish to assume that free will is indeed impossible, for others we should suppose that people do sometimes act freely. To assess the impact of genetic information on free will, it is important to consider the radical perspective of the free will dilemma, which challenges the notion of free will under any circumstances. This will save us from claiming that genetic information is a particular threat to free will because it would deprive us of something that, as we can see from the sceptical dilemma, we never had anyway. But if we are accurately to assess the impact of biomedical developments, it is also important to consider the more conventional perspective, which allows that there is a distinction to be drawn among the things we actually do, between those actions that are free and those that are not.

The conventional distinction between free and unfree behaviour treats free will as a default condition which may be compromised in various ways. Addictive behaviour is one sort of case. Certain people lack normal inhibitory mechanisms and so are unable to control their desires. Some people are unable properly to recognise or characterise the nature of some of their own actions. Here one thinks of cases of serious psychological impairment, but it is worth noting that there is also a version of this phenomenon that afflicts us all. Our actions invariably have effects we are in no position to identify: we do things unintentionally, and these are not done by our own free will. It is also worth emphasising how common are the cases both of loss of inhibitory mechanism and of ability properly to identify one's actions, as the problems of excessive drinking illustrate. On the assumption that heavy drinking is not itself always addictive behaviour, we have here also the important complication of cases where one freely chooses to make oneself unfree. And our free will may be compromised in other ways besides. Should the acquisition of genetic information be added to the list?

FOREKNOWLEDGE AND FREE WILL

Suppose that advances in genetics makes it possible to use information about an individual's genotype accurately to predict future behaviour. Would this sort of foreknowledge threaten free will? On the face of it, your knowledge of my future behaviour is irrelevant to my free will. Whether I act freely is a question of the kind of control I have over my actions: what you know about me seems irrelevant. Suppose that we are all lucky enough to enjoy free will. Now consider another situation, where our causal situation is unchanged, but there are invisible creatures observing us and discussing our performance. These creatures cannot interfere with us in any way, but they are extremely good at predicting what we will do. What difference does this make? If we had free will to start with, it is difficult to see how we would lose it in the face of this clever but passive audience.

Moreover, we already have a great deal of foreknowledge that requires no genetic information. This knowledge includes both general maxims of how almost anyone will act in a given sort of situation, and differentiated knowledge of how a particular individual is likely to behave, knowledge that is based on detailed acquaintance with his or her background, personality and previous behaviour, including verbal behaviour. If we grant that people have free will at present, we must take free will to be compatible with very considerable foreknowledge.

It is not as if free will would be more secure if we knew nothing about what people will do. For without foreknowledge, nothing like human society would be possible (Hume, 1748: section 8). If I knew *nothing* about how you will react to events, especially to what I say and do, we could share no projects, including the project of communication. Indeed if other people were completely unpredictable, we would not see them as agents at all. One reason for this is that we would find it impossible to attribute to them beliefs and desires.

Still, free will and foreknowledge might both be matters of degree, such that foreknowledge beyond a certain point would reduce free will. But as we have already observed, free will has to do with how actions are generated, not whether they were predicted in advance. I know some people better than others, and so I am better at predicting some people's behaviour than I am at predicting the behaviour of others; but I do not attribute more free will to those less familiar to me. Similarly, as I get to know someone better, I become better able to predict what they will do; but their free will is not thereby diminished. But these are not knock-down objections to the thought that too much foreknowledge might interfere with free will or its attribution. It might be that free will is compatible with a considerable range of foreknowledge, but not beyond a certain point. If my wife knew absolutely everything I was going to say (an achievement that sometimes seems within her grasp) perhaps she would find it difficult to see me as a free agent.

It is unclear whether even this extreme level of foreknowledge would really preclude free will; but even if it would, it is unlikely that this is the source of any special threat posed by genetic information. It is unlikely because of the enormously complicated interactions between genotype and environment in the aetiology of behaviour. Perhaps I underestimate the potential of the biotechnology, but it seems to me that while genetic information might show that Jones is substantially more likely to commit crimes of a certain sort than a randomly chosen individual, this kind of foreknowledge is very unlikely to rival the foreknowledge I could gain of Jones through close acquaintance, a source of foreknowledge that is not taken to pose any special threat to Jones's free will.

For foreknowledge to threaten free will, the knowledge itself must have an influence. For example, your knowledge of what I will or would do may cause you to treat me differently, and my knowledge of what I will or would do may feed back and influence my own behaviour. Could either of these influences create a world without free will? Perhaps the foreknowledge would interfere with the influence of desire, so that what we do would become independent of what we want. This is the difference between determinism and fatalism. In a deterministic world, every action has causes, but our desires may be among them; in a fatalistic world, desires play no role in the determination of action. What we do is determined, by environmental factors, by what others do to us and by features of ourselves such as our genotype, but what we want has nothing to do with it.

A fatalistic world is one without free will, but it is difficult to see how genetic foreknowledge could make the world like that. If desires were causes before genetic discoveries were made, they will continue to be such afterwards too. If desires really do now play a central causal role in action, as we all believe, this mechanism is so basic as to be impervious to fundamental alteration simply on the basis of new information.

GENES AND DETERMINISM

Perhaps we have been looking for the threat posed by genetic foreknowledge in the wrong place, because the threat comes not from the foreknowledge itself or from its consequences, but rather from something the knowledge would reveal. That is, the real worry may be not that genetic foreknowledge might destroy free will, but that it would provide compelling evidence that we never were free in the first place. The intuition is simply that precise foreknowledge would only be possible under a determinism that precludes free will. So as we gain that foreknowledge, we would also gain the knowledge that free will has always been an illusion.

If foreknowledge does indeed pose a threat to free will, this is I think the way the threat operates: foreknowledge provides evidence of determinism, and so evidence that we never had free will. Of course we know that genes do not by themselves completely determine behaviour in all its detail, since we know that identical twins do not behave in precisely the same way. But genetic discovery might make it appear increasingly likely that our behaviour is the output of a deterministic mechanism, in which genes play an important role. This does not however seem genuinely to pose a new threat. Recall the discussion at the start of this essay, that determinism in itself is no more or less of a threat to free will than its denial: either way we lose. This suggests that an increase in our knowledge is unlikely to reveal any special threat to free will simply by making it more likely that determinism holds: genetic information cannot pose a special threat to free will simply by exposing generic determinism.

There might however be a peculiar threat to free will revealed by genetic information, in virtue of the peculiar form of determinism it exposes. Even the most optimistic proponent of the compatibility of free will and determinism will admit that free will is not available in all deterministic worlds. One deterministic world without free will is a world without agents who enjoy beliefs and desires; another is a world in which beliefs and desires exist but have no influence on behaviour. Might genetic information reveal that we live in a world bereft of free will, not simply because it is deterministic, but because of the type of determinism it contains? (see Nuffield Council on Bioethics, 2002, ch. 12.)

In certain cases of genetic abnormality, this might be so. As has already been noted, we judge there to be loss of free will for example from addiction, lack of inhibition, and inability to recognise the nature of one's actions. Insofar as genetic information were reliably to predict such cases of diminished responsibility, it would reveal an absence of free will. But what about the normal cases? Even here one might worry that genetic information reveals a type of determinism incompatible with free will insofar as it reveals that we all have innate dispositions to do what we do. That these are dispositions suggests that they somehow bypass the mechanism of belief and desire that free action requires. That they are innate shows that the dispositions themselves could not have been chosen.

In my view, however, the innate dispositions that genetic information might reveal pose no special threat to free will. That one had a disposition to perform a certain action cannot by itself undermine the claim that this action was freely performed, since whatever people actually do we may say they were disposed to do. In many cases, we only discover these dispositions retrospectively, in light of the actions we see the agents perform, but that is not a difference in the disposition, only in how we discover it. Similar points apply even if we take 'disposition' to mean long-term personality trait. Long-term patterns in my behaviour reveal long-term dispositions, but these are not usually taken to show that I am bereft of free will. In any event, genetic information cannot be special in virtue of providing information about dispositions, since we regularly acquire that information by other means.

That leaves the innateness of dispositions, or of the genotype that is supposed to be their basis. Crudely put, you don't choose your genes, so insofar as your genes cause your actions, you don't choose your actions, so you are not free. But this argument is merely a variation on the free will dilemma. The causal history of our actions must extend back before we were born, and the fact that this history travels in part on genetic paths makes it no more or less a threat to our freedom. The causal history of our actions also extends outside our body, to the diverse environmental influences upon us, largely again beyond our control.

No good reason has been given for the claim that the genetic influence on behaviour should create a special threat to free will. The idea of genetic predictability makes vivid the thought that we have desires we did not choose; but given that our desires are not in any event determined by choice, it is difficult to see why the discovery that our genetic make-up plays a causal role should make any difference, so far as our free will is concerned.

Finally, perhaps the feeling of special threat comes not just from the fact that genes are innate, but that they are essences. An essence is a determinant of identity. Thus a piece of gold has both a shape and a chemical composition, but while it could survive a change of shape, it could not survive a change in composition, for then it would no longer be gold. Similarly, it has been claimed that people have essential properties, such as their parents. You could not have had different parents, in a way that you might, for example, have gone to a different school, for a child from different parents would literally be a different person (see Kripke, 1980: Lecture III). Now suppose that my genotype is one of my essential properties. In this case, there is a special and perhaps a specially deep way in which my genotype is beyond my control, something that goes beyond the way, for example, I cannot now change the school I went to. For although I myself can change neither, I might have gone to a different school, but could not have had a different genotype, while still being me.

If this were so, then genetic information would have the special status of revealing essences. This thought that genes are essential may be a source of the intuition that our free will is threatened by biological discovery. This is similar to the thought that genetic determination of action threatens free will because it shows my action is a consequence of my nature. What is particularly interesting about this form of the worry is that it suggests a kind of determination that goes beyond the causal determinism that has been our focus. For having a certain chemical composition does not simply cause something to be gold: it constitutes being gold. Similarly, on this causal–essentialist view, my genes not only causally determine my actions, but they are also constitutive of my identity.

We are now in deep metaphysical waters, but readers who have come this far will not be surprised to discover that I do not find here any special threat to our freedom. Indeed some might hold that it is reassuring to be told that one's actions flow from one's deepest nature, rather than from adventitious causes. But essentialism has little to do with free will. Perhaps I could not for example have had different parents while retaining my identity. This does not threaten my freedom. The same holds for my genes. If my genes causally determine my behaviour, then we have the familiar worries about determinism. But the additional claim that my genes are essential to my identity does not make the situation any worse. Presumably, one of my essential properties, if I have any, is being human: I could not be the same individual if I lost that property. But this hardly threatens my free will. The threat to free will comes from causal determination, not essentialism, and that threat is not new. It the old threat of generic determinism.

CONCLUSION: FROM CAUSES TO EFFECTS

This completes our whirlwind survey of the different ways in which advances in genetic science might be taken to threaten free will. The discussion has been framed by the philosopher's classic free will dilemma, an argument which highlights the double challenge posed to free will by determinism and by indeterminism. This places very high the hurdle that genetic information would have to clear if it is to pose a new and distinctive threat to our autonomy. I have considered the most plausible ways this might be thought to occur, whether by enabling us to predict people's behaviour, by undermining the role of desire in action, or by revealing causes of behaviour that are innate or even essential constituents of an individual's identity. My conclusion is that none of these factors make things any worse than the old dilemma already did. It can be deeply disturbing to be forced to face the ways in which determinism would make it true of all of our actions that we could not have done otherwise, and advances in genetic research may make it increasingly difficult for us to ignore this depressing fact. But even if the threat is thereby made vivid, it is not thereby made new.

That is the main moral of our discussion: advances in genetic knowledge will not in themselves pose a novel threat to free will. Another moral is that, when it comes to free will, we ought to worry less about causes and more about effects. On any plausible view of the world, our behaviour will be influenced by causes beyond our control. So if anything can make a difference, it is not the existence of such causes, but rather the kind of effects they have. When it comes to free will, what matters is our cognitive phenotype, not its genotypic source. Ordinary responsible behaviour and diverse pathologies that involve diminished responsibility may both have genetic bases, so the fact of genetic determination, insofar as it is a fact, will not explain the contrast. Of course if you want to alter effects, you will want to look back to causes that may provide you with effective handles. So the possibility of substantially increased powers of genetic intervention will give those concerned about human autonomy plenty to worry about. Genetic knowledge does not itself threaten free will, but what we do with that knowledge is another story.

ACKNOWLEDGEMENTS

I am grateful to John McMillan and the editors of this volume for constructive comments on this essay.

Part III **Neuroscience and the law**

6 Human action, neuroscience and the law

ALEXANDER McCALL SMITH

A cynical view regards the criminal law as a crude system of state-enforced sanctions designed to ensure social order: punishment follows the commission of a proscribed act. In one sense this is true: criminal law involves rules, the breaking of which results in punishment, and often rather crude punishment at that (a tyrant from an earlier age would certainly recognise a contemporary prison cell in a British prison, even if he might remark on its relative degree of comfort). This view of criminal law, however, is a very shallow one, and an anachronistic one too. Criminal justice is not purely concerned with the undiscriminating regulation of anti-social behaviour; a modern criminal justice system has far more sophisticated tasks than the mere punishment of those who break the law. These include the need to be imaginative in response to crime and to take into account the range of possible measures that can be invoked to deal with the offender. They also include the need to take into account the moral implications of the system and to ensure that the system is fair to those to whom it reacts. This is demanded not only by the moral imperative that the state should only punish the guilty, but by the pragmatic requirement that to be effective, the criminal justice system should enjoy a reasonable measure of popular support. A harsh system of criminal justice, which lacks the consent of the governed, will simply not work in an open and liberal society. All aspects of the system must be seen to be fair, consistent and morally defensible according to the prevailing sense of what is morally right. Of course there have been criminal justice systems which have been tainted with gross injustice and which

The New Brain Sciences: Perils and Prospects, ed. D. Rees and S. Rose.
Published by Cambridge University Press.

have nevertheless survived for longer or shorter periods. These systems have tended, in the long run, to be weak. They have left behind them nothing of value in humanity's general experience of law, other than a sour taste in the mouth of jurisprudence.

This need for moral acceptability has had two principal effects on the criminal law. One of these has been to foster an examination, the origins of which can be found in Mill and Bentham and which was particularly intense in the 1960s and 1970s, of the boundaries of the criminal law (Hart, 1963). The main thrust of this debate was that of establishing which areas of life should be within the reach of the criminal law and which should be matters of private choice (Feinberg, 1984–8). In the second half of the twentieth century, this led to a marked liberalisation of the law, particularly in the area of sexual conduct, where the criminal law progressively withdrew from attempts to regulate personal life. In criminal law theory, the concept of harm to others, rather than offence to others, became the main justification for criminal law's engagement, and the notion of the criminal law as an agent of state paternalism seemed increasingly unacceptable.

The other effect, and the one which principally concerns us here, was to encourage exploration of the extent to which the criminal liability could be made to reflect moral fault on the part of the offender. Of course the criminal law had purported to have been doing this for a very long time: the adage *actus non fit reus nisi mens sit rea* (the act is not criminal unless the mind is guilty) had long been cited by judges and legal writers as the central principle of criminal jurisprudence. What was required, then, was fine-tuning of the system in order to ensure that those who are convicted of criminal offences are, in fact, morally deserving of punishment on the grounds of their culpable state of mind. This process led to a debate between subjectivists, who believed that doctrines of *mens rea* should be made to do what they say they do – that is to ensure that only those who intend to do wrong should be punished – and the objectivists who favoured a response to harmful conduct rather than a fine examination of individual states of mind (Tur, 1993). A simple example of this difference of opinion has been the continuing debate about intention and about the extent to which we should be presumed to intend the natural and probable consequences of our acts. The subjectivist would say that we must only be held to account for that which we really intended (in other words, those consequences that we actually foresee). The objectivist would argue that it is reasonable to hold people accountable for those consequences which a reasonable person in their position *should have* foreseen. The current criminal law is something of a compromise between subjectivist and objectivist approaches. The criminal law does take into account subjective intention, but of necessity makes certain objective assumptions about what the subjective state of mind of the defendant was. There are several reasons why this has to be done. One of them is the fact that it will never be possible to ascertain mental states with complete confidence – people can, and do, conceal, lie and dissemble. In these circumstances we reach conclusions about what people meant to do by examining what they actually did – a matter of common sense to most people. But apart from that, even if we discount the difficulty of finding out what people actually meant, there is the further issue of the perceived need of the law to punish those who, even if they did not intend harm, have nonetheless behaved in a way which has caused harm to others. Here the law looks at the result and responds to that because there is a strong social demand to do so. An example of this would be the man who recklessly fails to ascertain whether a woman agrees to sexual contact. In some legal systems such a person might be convicted of rape even if he did not intend to commit rape. The justification for such an approach is that whether or not he intended to commit rape, he has, in fact, caused grave harm and outrage would be felt if he were not to be punished.

It is against the background of this tension between subjectivist and objectivist goals that one might consider the question of how far greater understanding in neuroscience might enable us to reformulate the criminal law analysis of human action. It is important to bear in mind that such an enquiry will have strictly limited implications for reform of the criminal law, precisely because of the conflicting considerations mentioned above and the fact that criminal justice ultimately responds to political decisions. The one thing that a stable system of criminal law will never do is to be over-generous with excuses (Stocker, 1999). A 'soft' system of criminal justice is a sitting target for political attack, and, in addition, it is likely to offend the public's sense of justice. A realistic aim, then, might be merely to point out how the insights of neuroscience may help us to understand the limitations of the criminal law's current view of human action. This is at least a beginning, and this understanding might be translated, at some stage in the future, into policy and assist reform at some stage in the future, even if the possibilities of reform at present may seem slight.

THE CRIMINAL LAW'S VIEW OF HUMAN ACTION

Criminal liability requires, as a first step, the conscious and willing commission by a free agent of an act which is proscribed by the law. As a second step, it requires that the agent should have a culpable mental state at the time of committing the act – a mental state which results from the making of a choice to act. We shall deal at this stage with the first step. This is usually expressed as the voluntary act rule: the criminal law will not impose liability unless the accused has committed a voluntary act. (Culpable omissions pose an obvious problem for this rule, but that is another issue.) There is no voluntary act if I am forced to act by overwhelming physical force – as where my arm is seized and forced through the motions of an assault on another – or where the act occurs without my volition – as in the case of a reflex action: if I am disturbed by a loud noise and drop something that I am holding. From the legal point of view, what is required then is a volition which accompanies or, more controversially, causes the physical action constituting the crime (Hart, 1968: 90).

The voluntary act requirement has been criticised on the grounds that the very notion of a volition is difficult to understand and does not adequately express what happens when we act in a particular way. These theoretical difficulties are perhaps overstated. There is a readily understandable difference between action which proceeds from an action plan conceived of by the agent and action which does not come from such a plan. For example, if A decides to fetch herself a glass of water and walks into the kitchen to so do, her act of walking into the kitchen and turning on the tap clearly bears a close relationship with the decision she has taken to do just this. Certainly she is not fetching the water inadvertently, nor indeed by accident. Similarly, if X, embroiled in an argument with Y, thinks that he is justified in striking him and then does so, the act of striking is likely to be associated with a decision or disposition to do so. At a more complicated level, the carrying out of a complicated fraud will proceed from a number of very specific decisions, and possibly after a great deal of internal weighing of risks and benefits to the actor. All of these mental events may reasonably be termed volitions if they are sufficiently closely associated with the execution of the corresponding action.

Thinking in terms of volition certainly helps us to make a distinction between actions which it seems right to refer to the actors as their acts – acts for which they should be accountable – and acts which we do not think should be morally attributable to the actor. Acts which proceed from our volition are therefore acts to which we stand in a particular relationship; they are acts which we feel have a connection with the continuing self which is the subject of moral responsibility. Acts which do not have this feature, are not *ours*; they are nothing to do with the self and it would be wrong to impose liability in respect of them. All this does, though, is to show that we must have experienced a particular mental state at the time of acting, a mental state which is referred to as a volition because it is composed of a *want* of the act in question. That seems simple enough, and accords with our everyday understanding of human action, but in reality the problem is only just beginning. The existence of a concept of the voluntary implies the existence of a concept of the involuntary, and we are then faced with the task of determining the boundary between the two. This boundary is a very significant one, as involuntariness will be a complete defence to criminal liability.

CONSCIOUSNESS AND THE VOLUNTARY

The legal philosopher H. L. A. Hart suggested that voluntary acts are those acts which we *take ourselves to be doing* (Hart, 1968). This is another way of saying that they are acts which we perform in a state of consciousness, or, rather, a state of consciousness which encompasses awareness of the act in question. It is possible to be conscious and still to be doing something of which one is unaware; an habitual tic or gesture of the hands is something that we might not be aware we are doing. We may know that we are prone to do it, and we might be aware of it when it is pointed out to us, but this is not incompatible with lack of awareness at other times. Consciousness, then, is a good candidate for the key ingredient of voluntariness, but it may be a minimum requirement rather than a sufficient requirement. If there are circumstances in which some features of consciousness are present and some are not, then voluntariness may be deemed absent on consciousness-impairment grounds.

Against this background, I will now consider in what circumstances consciousness may be incompatible with the performance of voluntary action. The law should find little difficulty with those cases where consciousness is self-evidently not present, as in a case of deep coma. Coma patients, however, do not perform acts of any legal significance – at least in the coma stage – and so this form of 'deep unconsciousness' is of no legal significance (other than in the context of the quite separate issue of the legal status of persons in the persistent vegetative state). Of course it is different when the person is on the edges of coma, as in the case of a diabetic in a state of hypoglycaemia, when there may well be a propensity to commit acts of violence, but such states are not states of coma as such and fit more into the impaired consciousness category (*R.* v. *Quick*, 1973).

If coma is the end of the spectrum, then the next point on the line between complete unconsciousness and full consciousness is the sleeping state. Sleep is accompanied by mental events, which can be detected by electroencephalogram (EEG) and which may be later remembered. There is a certain awareness of self in dreams, in that we undergo subjective experiences in which we believe ourselves to be actors; one might be tempted to call this a form of consciousness, but the sleeping state does not involve the degree of awareness of the position of self and the external world which is characteristic of consciousness as we generally understand it. For this reason it seems morally counter-intuitive to hold the somnambulist responsible for acts performed in the sleeping state, and not surprisingly this is the legal view (Shapiro and McCall Smith, 1997).

Criminal cases involving somnambulism are more common than might be imagined. Most somnambulistic activity is banal – consisting of opening drawers or cupboards or apparently pointless wandering – but somnambulists may occasionally engage in far more complex behaviour, including acts of directed violence against others. When confronted with such events, the courts have been prepared to accept the defence of automatism, which provides either for the acquittal of the accused or for his disposal as a mentally disordered offender. The choice of category of automatism depends on whether the automatic behaviour in question is considered to be the result of an 'internal factor' (effectively a brain disorder, or a disorder impacting on brain function) or whether it is the result of an external factor (such as a blow to the head, resulting in concussion). The crucial factor here is social danger and the possible need to detain those who might be dangerous in the future.

The recognition of a somnambulism defence is a concomitant of a requirement of voluntariness for criminal liability but the courts have still been anxious about the implications of acquitting defendants who have committed acts of serious violence, or even homicide. In one of the most remarkable cases, the Canadian case of *R.* v. *Parks*, the defendant was found to have driven a considerable distance, alighted from his car, and have killed his mother-in-law in a vicious assault – all while asleep (*R.* v. *Parks*, 1992). Parks was successful in his defence after psychiatric evidence was produced in support of his explanation of somnambulism; this led to some concern that the defence would be claimed in unmeritorious cases. Subsequent decisions of the Canadian courts, especially those involving defendants who have claimed to have committed sexual assaults somnambulistically, have underlined the tension that exists between the need of the criminal courts to balance social protection on the one hand with considerations of individual justice on the other (*R.* v. *Stone*, 1999).

Even if somnambulistic acts are placed firmly within the category of the involuntary, and therefore treated as proper cases for the application of the automatism defence, the boundaries between such acts and dissociative action will be considerably less clear. In one view, dissociation provides a better explanation for cases such as *R.* v. *Parks*; certainly dissociative states may trigger lengthy sequences of acts of this sort. But the problem for the law is that it prefers absolutes: a person is unconscious or he is not; a person has capacity or he has not; a person is liable or he is not, and so on. Dissociation is more subtle, in that although there is consciousness in the dissociative state, this consciousness is not normal consciousness and dissociation, which, even if controversial in some of its manifestations has come to be classified as a recognised psychiatric condition, may be at the root of inexplicable, out-of-character behaviour, and poses a particular challenge for the law. The person who acts in a state of dissociation is not unconscious in the conventional sense of the term; consciousness may be *impaired*, but whether that will be sufficient to justify acquittal is another matter. In general, legal systems are reluctant to allow this, and yet the sheer abnormality of dissociative behaviour demands at least some recognition from the criminal law. The difficulty lies in the understanding of the behaviour in question and in analysing it in terms of existing legal notions of action.

Several important cases illustrate the difficulty which the criminal law finds with dissociative behaviour. In the Canadian case of *R.* v. *Stone* the defendant was charged with the murder of his partner, whom he stabbed after experiencing a prolonged verbally abusive attack from her during a journey; a defence psychological report produced at the trial described the wife's comments as 'exceptionally cruel, psychologically sadistic, and profoundly rejecting'. While stopped at the side of the road, the defendant claimed that he felt a 'whooshing sensation' passing over him and when next opened his eyes he found that he had stabbed his wife numerous times and that she lay slumped beside him. The defendant then left the country for Mexico where he alleged that memories of the act of stabbing came back to him.

The facts of the Australian case of *R.* v. *Falconer* (1990) were similar, in that they too involved verbal abuse of a sexual nature. In this case the defendant was charged with the murder of her abusive husband, again after a prolonged and demoralising encounter between the two involving threatened sexual assault. After the husband revealed that he had sexually abused foster children whom the couple had been caring for, Mrs Falconer seized a loaded rifle and shot him. She had no recollection of the act of shooting.

The defendant's conduct in both these cases can be explained at various levels. One explanation – indeed the one which would occur most readily, and certainly to the average juror – would be that both Mr Stone and Mrs Falconer acted in a state of extreme anger when they had been subjected to extremely callous treatment. This interpretation might be couched in various ways: 'they struck out' or 'they snapped' would be demotic explanations. An alternative explanation, again an everyday one, would be that they both 'lost self-control' or were unable to control themselves. There is nothing wrong with such explanations: they express a feature of the behaviour in question which is quite relevant to the moral assessment of actor, namely, that the conduct lacked the element of self-control, with which people are normally expected to act (but for a feminist sociologist's perspective on this argument see Chapter 8, this volume). Anger overcomes self-control and the impulse to act violently is translated into action. Of course, this may not lead to a defence, as the law still expects from every person the same capacity to exercise self-control, even if it is the case that some people may have – for whatever reason, whether genetic or environmental – an inherent proneness to outbursts of temper (Horder, 1992).

Such explanations are understood by the criminal courts and are, in fact, written into a partial defence which is designed to deal with exactly such cases, the provocation defence. But a significant thing about them is that they do not point to any impairment of consciousness, focussing instead on a failure in the mechanisms of normal conscious action, and in particular of a deliberative process which weighs up alternatives and *controls* the response of the actor. It is questionable whether this model of what is actually happening in such cases actually fits the facts. Such conduct, surely, is better understood in terms of a model of behaviour which gives far less weight to the deliberative element of human action and which recognises that consciousness, control, intention and all the other constitutive elements of human conduct are not all-or-nothing absolutes but may be attenuated in such a way as to seriously affect responsibility. The issue for debate here is how can human behaviour of this sort be explained in a way that will allow a moral evaluation of conduct to take place.

It is instructive to imagine the position of Mr Stone and Mrs Falconer at the time at which they committed the acts which led to the charges of murder. Looking first at Mr Stone: he felt extreme humiliation at the hands of his wife, whose comments on his lack of sexual prowess and threats to deprive him of contact with his children could have been expected to undermine his confidence dramatically and, one might imagine, evoke rage. It is likely that he experienced a strong emotion of anger, and it was in this emotional state that he stabbed her so violently. Did he decide to stab her, though, or did he just do so without making any decision? This question may strike some as naive, and yet posing this question focusses our attention on those features of his action that count towards responsibility. If Mr Stone did not decide to do what he did, then is it a voluntary act on his part or is it something quite different? If he did not decide to stab his victim, then in a sense we can say that he did not *mean* to do it, and that his responsibility for the act is either negated or is appreciably diminished. It is very unlikely that Mr Stone engaged in any form of weighing of alternatives before he acted. In some instances this is exactly what we do – we deliberate on the relative attractions of different courses of action, and then make a choice as to one. More frequently, though, human acts are performed without a background of deliberation and evaluation. We describe this as 'acting impulsively' or 'acting spontaneously', and although we would not exclude the possibility of such action springing from a decision to act on the agent's part, such a decision is taken quickly.

Our understanding of what happens in the mind before we act may be imperfect, but it is clear that cognitive processes occur, situations are recognised and categorised, and connections are made which result in the performance of bodily acts. It is apparent that most of this takes place without our being aware of it, and that many of our responses are learned ones. Much behaviour, then, is not the decision of any deliberative and conscious decision to act, but results from the operation of the brain at a subconscious level. Driving a car, for example, or playing the piano involves learned responses which occur without our conscious direction once we have embarked on a course of action. Pianists do not make a deliberate effort to place their fingers on the correct keys to produce a chord which they read on the printed page; this happens without any deliberation on their part. Other physical movements fall into the same category. Walking involves the movement of limbs and the maintenance of balance through minor adjustments in posture. These are usually coordinated by the brain without any conscious effort on our part; they are the product of brain activity but this activity does not involve self-conscious states of mind.

If some human conduct – possibly a great deal – falls into this category of non-deliberative action, the question arises as to how far, from the point of view of moral and legal responsibility, such action may be excused. Moral and legal responsibility are not necessarily the same thing, even though in many cases they will coincide. Legal responsibility is affected by considerations of policy which may result in blame being attributed even where the agent could claim moral exculpation (Cane, 2002). This is the case with strict liability offences, or where an accused has made a genuine, but unreasonable mistake: in such cases the courts may decide that the interests of the victim or the interests of society require greater protection than the interests of the accused. In general, though, if we identify an act as being one for which moral responsibility is not appropriate, there is a presumption, at the least, that criminal law should follow suit and allow a defence. This presumption may be overturned, though, if there are interests that outweigh the argument for allowing a defence. These interests are potent ones: the principal one being that of protecting a general social interest by punishing individual transgression irrespective of the degree of guilt manifest in the individual action. We may therefore decide, and do so, that acts committed under provocation may nevertheless require to be punished, even if we recognise that the acts are out of character and, in some cases, entirely understandable in human terms. The criminal law serves a declaratory purpose: it identifies the forms of conduct that are unacceptable and then, in order to make an effective declaration about the seriousness of its purpose, it allows very little room for individual exculpation. The limits of this formal approach need to be explored, though, in order to ensure that there is a balance between the formal, declaratory punishment of objective wrongdoing and the requirements of individual human justice. One route to this balance is through the exploration of character theories of criminal liability, theories which, although unfashionable in criminal jurisprudence, have their adherents amongst moral philosophers. These theories may allow a place for understandings of human action which are based not on choice, but on notions of non-deliberative action.

Character means different things in different contexts (Flanagan and Rorty, 1993). Moral philosophers tend to regard character as the sum of the qualities in the moral self: the possession of virtues, for example, as manifest in attitudes and actions constitutes good character, while the possession of vices points the other way. For the psychologist, character is a tendency to behave in a particular way – a disposition which may be based on a wide range of constitutive elements and which is more likely to be described as a personality rather than a character. Traditionally, the criminal law has not been concerned with character, at least in relation to responsibility issues. However, character may be taken into account in sentencing, where in all but the strict 'tariff for crime' approaches, the accused's record and character may be considered. It is at this stage of the criminal justice process that the understanding of human behaviour as not being just about choices but about non-deliberative action may have some impact. If a person acts in a particular way because that is the way he or she is disposed to act – by virtue of character – then any decision about punishment should take this into account. It may be that deterrent punishment is required, but if punishment is to serve any other aims (such as rehabilitation, and therefore, indirectly, social protection) then the extent to which character, rather than choice, is the source of the offending behaviour might be investigated. This may result in the asking of fundamental questions about sentencing alternatives, the provision of therapy in appropriate cases (for example in the case of sexual offenders where crude retributive punishment is very unlikely to affect recidivism rates), and about the whole nature of the criminal problem. Pinning all one's hope on persuading people to make pro-social choices is likely to distract effort from confronting the deeper causes of crime, which are multifactorial but which in many cases will include an element of individual and social pathology.

Of course a standard objection to any such attempt to widen our concern to embrace such explanations of behaviour is that it leads inexorably to a medicalisation of conduct and to the weakening of individual responsibility. If we find excuses for our behaviour in our personal psychology, then we dilute the necessary measure of blame that we need for basic social order. This is not necessarily so. Responsibility may remain intact while at the same time we understand why it is that a person becomes inclined to offend. The insight which neuroscience may provide here is that such behaviours as aggression may be learned responses which may be avoided through various forms of social intervention. That, of course, is a Herculean task, especially in a market-based vision of society in which communal values tend to be eclipsed by individual, consumerist goals. We may suspect that the marketing of aggressive computer games may heighten violent behaviour in children, or at least a belief in children that violent behaviour is acceptable or normal. But to control such devices is virtually impossible in a culture in which censorship is difficult, if not impossible.

BEHAVIOURAL GENETICS AND PERSONAL RESPONSIBILITY

Even if one were to classify some forms of non-deliberative or 'quasi-involuntary' acts as acts for which responsibility might to some degree be attenuated, any such process would be faced with the objection that the factor which leads to such behaviour – possibly character or personality traits – is to an extent within the control of the actor and may therefore be something which he or she might be expected to ameliorate or at least control. Such factors do not therefore become completely deterministic. Neuroscience, then, may explain how conduct occurs, but does not necessarily completely exclude a reviewing, morally responsive self, which can weigh up its past behaviour and make moral effort to avoid events or situations that trigger problematic behaviour. However, if people were to be shown to be 'hard-wired' in such a way that the learning of new behaviour or new responses was impossible, then the determinist position becomes more beguiling and it may be rather more difficult to defend individual responsibility. We may still reject determinism in these circumstances, but our rejection may be entirely pragmatic, and based on social need rather than on moral coherence.

A major deterministic challenge today comes from behavioural genetics, a controversial area of genetics which, in the view of some, is potentially open to objectionable abuse (Nuffield Council on Bioethics, 2002). The central premise of behavioural genetics is that human behaviour is influenced by the genetic characteristics of the individual, the route of this influence being either through body chemistry or through variations in brain structure. The identification of the precise mechanism whereby this influence occurs will be a difficult task, and it will be even more difficult to identify the specific genes involved, given that the overall effect is likely to be produced by the interaction of a number of genes with each other and with the environment.

In spite of these difficulties, behavioural geneticists have been able to produce sufficient evidence to justify serious thought being given to the legal and ethical issues of this field of activity. Research in human behavioural genetics has been concentrated so far on three main areas: intelligence, sexual orientation and so-called anti-social behaviour, the last of these being most relevant to issues of responsibility. Insofar as the consequences of gene action necessarily affect all aspects of human behaviour it is clear that genetic differences between individuals may contribute to differences in the way an individual responds to particular social challenges, potentially resulting in inappropriate or socially unacceptable behaviour. There have not been many studies linking specific genes to particular forms of behaviour, but one in particular has attracted considerable attention. In this study, researchers focussed on three generations of a widely dispersed family in the Netherlands in which a small number of male members were claimed to have been involved in various forms of unacceptable behaviour including arson, rape and 'having a violent temper'. Eight such 'aggressive' male members of the family were found to have a defect in a gene on their X chromosome which was responsible for the production of monoamine oxidase, an enzyme which affects serotonin levels (Brunner *et al.*, 1993). This line of enquiry was later pursued in a study of 500 male children, which suggested that those children who had a gene which resulted in low levels of monoamine oxidase were more likely to behave violently in adult life, but only if they were exposed to maltreatment and abuse in childhood (Caspi *et al.*, 2002). This points to the combined operation of genetic and environmental factors, and neatly illustrates the pitfalls in assuming that the identification of a genetic factor with a particular form of behaviour constitutes the sole explanation for that behaviour.

If more evidence becomes available of a link between genetic characteristics and behaviour, will this seriously challenge our existing notions of responsibility, and will it be evidence of the sort that can be taken into account in the criminal courts? This question is, to an extent, the same as the responsibility question prompted by advances in the neurosciences, the only difference being that in the case of genetic factors one is going one stage further back in finding the cause of behaviour. Moreover, these questions are not entirely novel. Lombrosan theory, which has existed as an enthusiasm in criminology from the nineteenth century onwards, has looked for the biological basis of criminal behaviour, from both the point of view of explanation and prediction. Lombroso and subsequent phrenologists concentrated on the shape of the head and on the face, and this broadened to an interest in physique. In all cases the objective was to link anti-social conduct with a particular physical type. Subsequently the emphasis switched from the phenotype to the genotype, with the interest in the alleged link between the abnormal XYY chromosomal constitution and violent behaviour being a noted and flawed example of this approach in the mid 1960s (Jacobs *et al.*, 1965). Contemporary monoamine oxidase studies are examples of what is essentially the same quest: to identify a biochemical or genetic basis for criminal behaviour. They differ substantially from their antecedents in method and rigour, but the same question might be asked of all those who engage in this quest: what are the implications of finding a close correlation? Can 'aggressive or violent behaviour' be defined in a manner that makes sense as a reproducible phenotype that can be studied genetically? If so, and if there are found to be genes that are contributory to the behaviour, then will this be an exculpatory factor, or will it remain a predictive or explanatory tool?

There is useful comparison to be found in the condition described as psychopathy and in the effect that this condition has on responsibility. The person with a psychopathic personality disorder has not chosen to be afflicted by a disorder. Whatever the aetiology of personality disorders may be (and there is a range of theories, including theories of neurological abnormality), nobody suggests that the individual can control whether or not the personality disorder develops in childhood. We can conclude, then, that it would be quite wrong to blame the psychopath for his psychopathic personality disorder, or, in other words, to blame him for being a psychopath. The absence of volitional choice does not preclude responsibility for the acts which that person performs, unless the act in question can be demonstrated to have been one which was itself entirely determined by the personality disorder. Take the case of an anti-social form of sexual behaviour, paedophilia being a useful example here because of the strong social condemnation of this form of behaviour. The paedophile is not held responsible for the fact that certain stimuli are sexually attractive to him. What he is held responsible for are any actions on his part which result from his paedophiliac disposition. Is this unjust? The paedophile seeking to refute responsibility might argue along several lines. He might say that it is unreasonable to expect anybody to deny himself sexually altogether, and that it is therefore unreasonable to expect an individual to resist the promptings of such a powerful drive. This argument, however, is likely to be rejected, on the grounds that the gain to the individual in this case (sexual fulfilment) is far outweighed by the harm caused to the victim. Sexual drives are viewed as resistible, even if this may involve considerable sacrifice to the person called upon to resist.

A second line of argument, and one which would be more likely to attract some support, is based on the notion that the paedophile's condition constitutes an illness, and therefore has the same excusing status as does any illness. The fact that paedophilia and other sexual paraphilias are listed in psychiatric diagnostic manuals is important here, as is the fact that the criteria which are generally used to qualify a condition as an illness may be present in paedophilia. Of course this does not resolve the matter either legally or morally, because the presence of an illness will not of itself be regarded as being grounds for exculpation in respect of any particular act. What has to be shown is that the act is so closely linked to the illness as to be explained or produced by it, and, furthermore, that the illness puts the act either beyond the understanding of the actor or beyond his control.

Mental illness is generally held to a morally exculpatory factor (Reznek, 1998). This is a very strong moral intuition, but further examination of the basis of this moral intuition suggests that what exculpates in psychiatric illness is the fact that the actor's understanding of what he is doing, of its impact, or of his environment is so distorted by the illness that his action cannot be the subject of moral assessment. Similarly, if control is the issue, the person who acts under the influence of an illness may not have had sufficient ability to refrain from behaving in a way which they may even understand to be wrong or the subject of moral disapproval. Both of these tests involve a reasonably high threshold of non-responsibility, and would not, for example, preclude responsibility in the case of many low-level psychiatric conditions, even if these conditions have some impact on behaviour. By contrast, a florid psychotic condition is much more likely to be considered a complete excuse.

In the legal context, tests for what is known as the insanity defence, which leads to the total exculpation of the accused, tend to require that there should have been a fairly substantial disturbance in cognitive or volitional capacities before the defence will be admitted (McAuley, 1993). It will not be sufficient, then, to demonstrate mental abnormality or disorder: it will need to be shown that the condition in question was sufficiently serious to negate understanding of the nature and implications of action, or, in the case of some volition-based tests, to have removed the accused person's capacity to control impulses. The relatively narrow interpretation which has been given to these tests in many systems of jurisprudence is based on concern that 'soft excuses' will be allowed and that the insanity defence will be claimed by those who are looking for an easy way out of a serious charge. In the United States, for example, there has been considerable political hostility to the insanity defence in some states, where it has been felt that this defence has been abused by psychiatrists who are willing to stretch psychiatric explanations of behaviour too far (Mackay, 1996).

It would be very difficult to see genetic explanations satisfying the stringent criteria which have been established for the insanity defence. Even if it could be demonstrated that a person has a gene which results in a demonstrable abnormality (for example, low serotonin levels), this will be unlikely to satisfy specific legal criteria for cognitive or volitional incapacity, especially as the behavioural consequences of such an abnormality can be very varied. This does not mean, though, that there might be no role for such evidence in relation to the more flexible defence of diminished responsibility. This defence, which is not a complete defence but which has the effect of reducing a charge of murder to one of manslaughter (in English law at least), is considerably more flexible in terms of the criteria which must be satisfied. Depressive illness, for example, may form the basis of a defence of diminished responsibility, even if it cannot be shown that the effect of the depression was such as to prevent the accused from understanding the nature of his acts or from complying with social and legal norms. There must, however, be a recognised psychiatric condition, and this means that the genetic condition must be linked to an illness; an asymptomatic condition which merely created behavioural traits would not be sufficient.

From both a moral and a legal point of view, the fact that a person has acted in a particular way because of a characteristic of the brain or of the genotype, may not exculpate but may lead us to mitigate blameworthiness to some extent (Wasserman, 2001). Returning to the psychopath, or indeed to the paedophile, such persons may behave anti-socially, and may be held responsible for what they do, but may still deserve some pity at least on the grounds that the starting point for their choices is so much more difficult than it is for those who are not afflicted by such conditions. The inactive paedophile or the inactive psychopath (if the latter is a credible term) does not deserve our censure; if either acts in a harmful way, then we may legitimately censure, but at the same time we might acknowledge that it must have been very much more difficult for that person to comply with social expectations. This is not the same thing as exculpation, but it does suggest that less punishment may be exacted or that, indeed, the conduct in question could be approached from a social protection viewpoint rather than punitively. It may also suggest psychiatric disposal, if that is feasible and available.

In the final analysis, responsibility survives the insights of neuroscience and behavioural genetics. This is partly because these insights do not completely destroy choice and the concept of deliberative action. They weaken it, certainly, in that they suggest that at the boundaries, choice may be less central than we would like to imagine, and that some forms of behaviour are much less deliberative than the traditional legal analysis of human action suggests. The appropriate response to any such challenge is an improvement in our response to crime, and this involves a delicate balancing of maintaining individual responsibility with more imaginative, therapeutic options for offenders, including, crucially, much greater commitment to early intervention. It also calls for a re-examination of how social attitudes to violence, aggressive and self-seeking behaviour can result in patterns of learned behaviour that will eventually contribute vastly to the social and financial cost of offending behaviour. That will require moral commitment to shared values which modern pluralistic, market-driven visions of society might find it difficult to accommodate.

7 Responsibility and the law

STEPHEN SEDLEY

When a defendant claims in an English court that his offence was not murder but manslaughter by reason of diminished responsibility, the judge has to explain to the jury that:

> Where a person kills or is party to the killing of another, he shall not be convicted of murder if he was suffering from such abnormality of mind (whether arising from a condition of arrested or retarded development of mind or any inherent causes or induced by disease or injury) as substantially impaired his mental responsibility for his acts or omissions in doing or being a party to the killing.

This provision was introduced into the law of England and Wales by Act of Parliament in 1957. It had by then been part of the judge-made law of Scotland for many years. Both jurisdictions had also for centuries reduced murder to manslaughter in cases of provocation. Provocation too was recognised by Parliament in 1957, but its classic formulation, which judges are still required to use, was given to a jury by Lord Devlin as a trial judge in 1949:

> Provocation is some act, or series of acts, done by the dead man to the accused which would cause in any reasonable person, and actually causes in the accused, a sudden and temporary loss of self-control, rendering the accused so subject to passion as to make him or her for the moment not master of his mind.

The New Brain Sciences: Perils and Prospects, ed. D. Rees and S. Rose.
Published by Cambridge University Press.

Both in terms of oral prose and in terms of lawgiving, this is as good as it gets. Like all judges of the Queen's Bench division of the High Court, I spent years – six in my case – trying homicide cases on circuit. I became and remain an admirer of the jury system, largely because of the rapidity with which twelve lay people were generally able to grasp and apply to a live problem before them principles of law which, as abstractions, continue to baffle undergraduates after three years' study. But one watched the jurors' faces with increasing despair as they listened to the direction on the difference between manslaughter and murder.

Deciding whether the accused had lost his self-control was usually straightforward, but it was only the start of the problem. The jury then had to postulate a reasonable person in the same situation and to decide whether he or she too would have lost it. Refinement over the years of the Devlin formula has – sensibly – required the notional rational comparator to be invested with some of the accused's own characteristics: a visible deformity, for example, so that the question becomes how a reasonable person with the accused's deformity would have reacted to being called a cripple; or a chronically battered woman facing one beating too many. The reasonable person who alone could be legitimately provoked has metamorphosed in many cases into a person whose reason is besieged by enough misfortunes to drive many reasonable people to distraction.

In diminished responsibility cases, too, one's judicial heart sank. Psychiatrists had to enter the witness box and testify about the accused's mentality in a vocabulary that no modern psychiatrist (and probably few in 1957 either) would use. They had in particular to address something described by Parliament as impaired mental responsibility. One felt oneself, alongside the jury, to be peering into a very deep pool indeed with very little help about what was to be found there.

There are two major characteristics shared by most killers to which, however, the law shuts its eyes: one is a violent temper with a short fuse; the other is a skinful of drink. Logically the jury should be considering whether a reasonable person with *these* characteristics would have lost his self-control; but to do so would be to ensure that practically every killer who ran a defence of provocation was acquitted of murder. It would also ring-fence the very sources of violent behaviour which the law hopes to suppress. So the law, as a matter of policy, excludes them from the jury's consideration unless they were themselves the topic of the victim's provocative conduct or unless they are real pointers to an abnormality of mind.

But please do not think that this chapter is going to be a disquisition on the English law of homicide. I have barely scratched the surface of that. What I am trying to illustrate is that the law on human responsibility for acts which harm others is a set of historic and moral compromises. The schematic and at times arbitrary defences to murder, where conviction was until 1957 a passage to the gallows, owe much to this. Most lawyers now recognise that there needs to be a single offence of unlawful homicide, the sentence reflecting the degree of culpability, and that the present quantum leap between the penalty for murder and that for manslaughter is both absurd and unjust.

Not even this much-needed reform, however, will penetrate far into the philosophical and biological thickets which beset the law as they beset the rest of life. Judges do not know – and it is a comfort that scientists do not know either – to what extent a flare-up of temper under minor or imaginary provocation is genetically or constitutionally determined. Judges cannot say – and it is a comfort that philosophers cannot agree either – to what extent a person who drinks himself into a state of aggression should be held responsible for all the consequences. Instead, because the law is for ever trapped between the need for certainty and the need for adaptability, we draw lines in the sand. We hold that because drinking is a voluntary act its consequences are likewise acts of volition. As a proposition of human pathology the conclusion does not begin to follow from the premise, but as a matter of social control it has to. If the law allowed drunkenness to stand between the individual and his responsibility for what he does, we would see a genuine rent in the social fabric.

We reason out this issue, whether we admit it or not, from consequences. Law, unlike science, can do that without being dismissed as unprincipled, because law is legitimately concerned with consequences. The difficulty it faces is that if consequences are all it looks to, principle and consistency, which a good legal system should regard as more important, will be sacrificed on the altar of short-term outcomes.

This is why it was initially tempting to stick in this chapter to homicide and its variants, the popular high ground of legal responsibility. But few people deliberately kill. Far more harm others by stupidity or inadvertence, and by doing so attract not a criminal but a civil liability to pay compensatory damages for sometimes unexpectedly dramatic consequences. They may be motorist, employer, fellow worker, doctor, social worker, teacher: as I list them your assent, I suspect, falters. It's one thing to say that a driver who fails to keep a proper lookout for pedestrians ought to pay for the consequences. It's another to say the same of a teacher who fails to pick up dyslexia in a slow reader.

Yet the leg broken by the careless driver may heal without further ill effects; the child mistakenly written off as stupid and idle at eight may as a direct consequence have become an illiterate criminal by the age of fourteen. The prisons are full of them: half the prison population has a literacy age of eight or less; two-thirds a similar numeracy age. Some are so damaged by emotional, sexual and physical abuse, exclusion from school, neglect in care and a dozen other breaches by adult society of the obligation to afford them some self-esteem and a fair chance in life, as to be driven by little other than survival techniques, even if these are also (like drug abuse) techniques of self-destruction. Many of these offenders are of low mental capacity; but many too are alert and intelligent. For all but a few of them (most typically among those few the drug dealer who is not a user), crime has been less a choice than a course predetermined by a mass of social, educational and economic factors from which it would have required both luck and determination to escape. One might have said, at least philosophically, that they are barely responsible for anything they do – but outside the law of homicide, mental and emotional handicap counts for nothing until it reaches a level of mental disorder which signposts a secure hospital in lieu of prison. Once again the law, for fear of what would otherwise follow, holds them all answerable for what they have done to society.

But what has society done to them, and who is to answer for it? It has for a long time been a principle of English law that a person, whether or not a state official, who inflicts foreseeable harm on another by neglecting to take reasonable care is answerable for it in damages. In recent years the applicability of this unexceptionable principle has been put under stress by endeavours to extend it from the classic categories which I listed first to the much more troubling categories I went on to mention. Doctors perhaps are the class which articulates the two. We have come for the most part to accept that a doctor, while in no way a guarantor of correct diagnosis or cure, has to bring a professionally determined minimum standard of competence to these tasks and will be liable in negligence if a failure to do so causes harm to the patient. Subject to a generous margin of judgement which the motorist, for good reasons, does not enjoy, the logic and justice of subjecting both to the same principle of liability is apparent. Why not then teachers and social workers?

Once again, consequences loom large among the reasons for drawing a line. Defensive social work and teaching will come in the wake of defensive medicine. Stretched social work and education budgets will collapse under the strain of fighting or meeting such claims. The claims will be made, moreover, not for the generality of victims but for the handful who find their way to the offices of aggressive and determined lawyers. If they succeed they will require the courts to speculate wildly on the value of what might have been achieved but for the failure.

These and other considerations led the Appellate Committee of the House of Lords during the 1990s to shut the door on a series of claims that were going to extend liability in negligence to the police and local authorities for errors of judgement which, if the claimants were right, went beyond mere mistakes and amounted to gross neglect of their duties. Those duties, said the law lords, were owed by officials to the public or the state, not to individuals. Responsibility, in other words, existed, but it was to be enforced, if at all, by the disciplinary and other mechanisms of the public services themselves, not by making those services answerable to the individual casualties of their officers' errors.

The European Court of Human Rights disagreed. In two cases in particular, each involving a gross neglect by public authorities of an obvious danger to individuals' life and health, the Strasbourg court held that the courts could not confer an a priori immunity from actions for damages without negating the right given by the Convention to redress in each member state's courts for violations of the substantive rights to life and to freedom from inhuman or degrading treatment. We have had in consequence to return to the juridical drawing board. We are not required to abandon the notion that professionals are entitled to room to make mistakes or to refrain from stifling proceedings which cannot surmount that hurdle; but we are required not to let it grow into a doctrine of official immunity.

It is in the measure by which the state enforces the responsibility of people to one another, as much that of people to society, that the courts have to make and remake value judgements that travel well beyond the jurisprudential. They have had repeatedly to grapple with theories of causation which have troubled philosophers and scientists for millennia, yet with only the practical toolbox of jobbing lawyers to assist them.

If I ride my bicycle through a red traffic light and cause a motorist to veer, am I responsible not only for his dented vehicle and black eye but for the collapse of the limited company through whose shop window he goes, wrecking the promotional sale that was going to save it from insolvency? If so, am I also responsible to the employees who in consequence lose their jobs? If I am, am I also responsible to their families for the hardship it causes them? A logician would say that I have caused all this and more to happen; and I suppose a chaos theorist would add that its very unpredictability was predictable. The law agrees to an extent: it does not limit the wrongdoer's liability to what the wrongdoer could have anticipated: it accepts that if harm was foreseeable, it's the wrongdoer who must bear the risk of unforeseen as well as obvious consequences: but it has finally to recoil from its own logic and draw lines. Again, its reason for doing so is the consequences of not doing so.

Law is a discrete discipline which lays few claims to scientific or philosophical integrity. But the virtue (as lawyers think it to be) is also a vice, because it limits the range of our inquiry. We do not know what we shall do if aggression turns out one day to be biologically determined, or if pharmacological or genetic intervention becomes able to produce results which fines and imprisonment do not achieve. We listen in certain cases to the evidence of scientists about, for example, the effect of testosterone treatment on sexual behaviour; but endocrinology's own account of this has swung through major cycles in the last twenty-five years, leaving the courts bemused. We listen in other cases to scientific witnesses on both sides whose reasons for disagreeing with each other are sometimes, no doubt, venal but are on other occasions the product of genuine epistemological rifts within their own discipline or possibly of ideological differences between individual professors of it. We are also from time to time treated to solemn excursions into junk science.

It is not, in other words, as if there were waiting out there a fund of facts about human responsibility on which the law simply declines for its own obscurantist reasons to draw. It is rather that in the absence of a confirmed or at least a compelling aetiology of human conduct, the law adopts certain scientific or quasi-scientific baselines – mental abnormality, automatism, self-control – but goes on to construct its own paradigms of human responsibility from its experience of two chief things: the ambit of human frailty and the social consequences of irresponsibility. I don't say for a moment that the law has got it right, but in that, at least, we are not alone: a little more than a century ago Lord Kelvin, the president of the learned institution which hosted this valuable symposium, announced that radio had no future, that heavier-than-air flying machines were impossible and that X-rays would prove to be a hoax. We shall continue, I hope, to learn from one another.

8 Programmed or licensed to kill? The new biology of femicide

LORRAINE RADFORD

This chapter considers the contributions that new biological research in the areas of behavioural genetics, brain anatomy and neurochemistry may make to our understanding of and responses to male violence against women as a public health problem. It also tests recent claims that the new biology can seek out and treat the dangerously femicidal and violence-prone.

FEMINISM AND THE PROBLEM OF FEMICIDE

'Femicide' is the 'misogynous killing of women by men' (Radford and Russell, 1992). The term points at the politics of woman killing at both the individual level and that of governance. It covers the many forms it takes – marital, stranger and serial killings as well as racist and homophobic killings of women. This said, however, most of the research discussed in this chapter homes in on either marital femicide (wife or partner killing, often called uxoricide) or the much rarer 'lust driven' serial killing of women by men. The rate of general homicide overall in England and Wales is relatively low – 15 per million of the population, with men showing higher rates of victimisation than women at all ages. On average 70% of homicide victims in England and Wales are males and 30% are females. Women less frequently kill and they are much more likely than are men to be killed by a partner. On average in England and Wales, two women are killed by male partners or ex-partners each week (*Criminal Statistics*, 2001).

The New Brain Sciences: Perils and Prospects, ed. D. Rees and S. Rose.
Published by Cambridge University Press.

Newspaper coverage of femicide is prolific and often voyeuristic, portraying male perpetrators as either villain (*Guardian*, 2002):

> A man was yesterday sentenced to life for murdering his wife in 1978, after a jury was told by his daughter how she witnessed the killing as a child but had only recently remembered what she had seen . . . The prosecution described (Bowman) as a violent man who raged a 'reign of terror' against his family and attacked his 44 year old wife because she had threatened to leave him. He pushed her violently, knocking her unconscious when she hit her head on the fireplace. He then force fed her with alcohol and drugs before finally strangling her.

or as a misunderstood 'ordinary dad' driven to commit a grisly event in an otherwise ordinary suburban home (*Guardian*, 2001):

> A terrified seven year old girl raised the alarm as her policeman father bludgeoned to death her mother and younger brothers before hanging himself in their Kent home, it emerged yesterday. Karl Bluestone, 36, a police constable, attacked his wife Jill, 31, and their four children with a claw hammer in a frenzied assault. His daughter Jessica, who suffered minor injuries, was able to run to neighbours to raise the alarm . . . Yesterday, residents left flowers at the semi-detached 1960s house as forensic officers gathered evidence inside . . . Neighbours on the estate, where PC Bluestone had grown up, were shocked that such a popular and well-respected figure could kill the children he adored.

PC Bluestone was alleged to have killed his wife and children because he found out she had an affair and was going to leave him. It was subsequently discovered that he had a history of domestic violence towards his wife and had previously been arrested for 'accidentally' injuring one of the children.

Challenges to male authority, sexual jealousy and the woman's actual or threatened separation are very common factors in domestic violence and in domestic killings. Three out of five homicides of women studied in Australia resulted from a 'domestic altercation'; 40% of these women were killed at the point of separation or after leaving (Mouzos, 1999). The failure of the police and other agencies to act against domestic violence is frequently the real-life tragedy behind murder–suicide cases like Karl Bluestone's.

It is the normalisation of violence against women and its social construction as a less serious crime that has occupied feminist writings (Lees, 1997). One strand, that of radical feminism, views all acts of violence from men to women, from femicide, rape, woman battering, prostitution and pornography to wolf whistles, as part of a 'continuum of sexual violence'. The continuum secures and maintains the relations of male supremacy and female subordination that are central to a heterosexual patriarchal social order. On the continuum 'typical' and 'aberrant' behaviour for males blend into one another (Kelly, 1988). The alternative social constructionist perspective explores the victim blaming and 'iatrogenic potential' of the law; the tendency of courts, like medicine, to makes things worse in the process of striving for a solution. Justice is said to be biased in favour of men who abuse women and victims themselves are put on trial. The violence is individualised, the courts respond to a single event and rarely take sufficient account of the history of domestic violence. Lees (1997) argues that a woman's sexual reputation plays an important part in whether or not courts will regard her as a 'deserving' or a 'non-deserving' victim.

Inequality between men and women makes women 'suitable' targets for men's violence and can make it much harder for the woman to move away. Worldwide, the highest rates of violence to women are found in countries where women and girls have low status in comparison to men and boys. Access to alternative sources of support and accommodation without doubt makes it more possible for women to leave abusers whilst dwindling options and lack of alternatives will make it harder. Greater equality with men can cushion women from violence. Looking at homicide data for 158 cities in the USA, Vieraitis and Williams (2002) concluded that homicides of women by male partners varied between states in association with variations in women's absolute status and, for white women only, their status relative to men. These variations in femicide rates, power and inequality are intriguing and warrant further research. I will show later how this is an area that has also gained attention within the new biological perspectives on violence.

Although marital femicide is a comparatively rare event in England and Wales, fear of being killed, particularly by strangers, plays a part in women's everyday lives, their experiences of safety and danger and their dependence upon known men as protectors As Lynne Segal (2003: 217) notes:

> When a flasher jumps out at a woman, or a voyeur lurks at our window, he is usually not a rapist or killer. But he just might be. His actions certainly serve to make the world feel unsafe for women.

Critics have argued that this promotes the conclusion that all women are victims and all men abusers. Although there has been a general opening out of debates about gender and violence to take into account women's violence and violence to men there is little research on why some men do and others do not abuse and kill. Not all men are abusers but it is hard for women to know who is and who isn't. It remains that the threat of violence, even when it does not occur, shapes women's behaviour.

To sum up this section we can conclude feminist perspectives have shifted attention on to the politics of femicide, at the level of individuals – the women and children who are killed and the men who kill them – and the wider level of governance, how the violence in legal, scientific and cultural discursive practices is made either 'normal' or pathological. Claims that biology sits at the root of violent behaviour are treated warily as part and parcel of the individualisation of the problem of femicide and the tendency to blame women for being victims of their own demise.

THE NEW BIOLOGY OF VIOLENCE

There is a long history within criminology of the idea that criminality, deviance and aggression can be explained on the basis of biology. In the nineteenth century Cesare Lombroso observed physical characteristics, skull sizes, foreheads and jaw shapes to support his argument that criminals were 'atavistic', physiological and moral throwbacks in human development. More recently there has been a history of research on twins and adopted children to see how nature, nurture or environment influence aggression and violence. There has also been the – now discredited – research on men imprisoned for crimes of violence with chromosomal abnormalities such as XYY.

As a subject discipline, however, criminology has favoured sociological perspectives on crime and deviance over biological. Up until very recently, criminology has dismissed biological perspectives as 'positivism' with dangerous implications. But new biological approaches differ from the deterministic earlier versions as they are much more likely to acknowledge that the factors influencing violence and aggression are complex. The key point argued is that genetic, neurophysiological and neurochemical factors, combined with upbringing and social and environmental variables, can create within an individual a propensity towards violent behaviour. Identifying this propensity has implications for how a society is able to manage the risk of violence and homicidal behaviour.

Violence and homicide have been explored in the context of evolutionary psychology, behavioural genetics, neuroanatomy and neurochemistry. I will summarise the key arguments within each approach before moving on to discuss their possible impact upon the prevention of violence against women.

Evolutionary psychology

Evolutionary psychologists tend to be critical of feminist 'bio-phobia' arguing that their critiques have been based more on politics and ideology than on sound science (Thornhill and Palmer, 2000a). Up to a point this criticism is justified. There is very little research into femicide and what there is has been mostly produced in the USA where homicide rates and access to weapons which influence these are considerably higher. Evolutionary psychologists share with feminism an interest in the gendered nature of homicide focussing on uxoricide (wife or partner killing) and familicidal massacres (where the mother and children are killed by the father). Controversially, evolutionary psychologists use allegedly Darwinian ideas about natural selection to argue that men's tendency to rape or kill women and children is an evolved and adaptive trait (Thornhill and Palmer, 2000a, 2000b; Wilson and Daly, 1999; for critiques, see for example Patrick Bateson's chapter in this volume and Rose and Rose, 2000). Thornhill and Palmer argue that rape may be an adaptive trait for men who cannot otherwise find a sexual partner and hence reproduce. Although killing a mate and her children would seem not to be in a killer's genetic interests evolutionary psychologists such as Daly and Wilson argue that it has continued as a worldwide aspect of human male behaviour because it is a by-product (epiphenomenon) of the evolved psychology of men, uncertain as to whether they were the true genetic parent of their mate's child. Evolutionary psychologists also make the general argument that violence and aggression in males aided genetic fitness in the savannah grasslands of Stone Age hunter–gatherer societies. Males who succeeded would be those who could direct their aggression to catching and killing food, competing with other males for mates and protecting women and children to ensure successful reproduction of their progeny. Women likewise, it is argued, would chose dominant, wealthier and violent males as mates to be the best providers and as 'mate guards' or protectors. Violence towards female partners and children thus has its roots in the human male's need to fight off rivals and in his fear of cuckoldry. According to Wilson and Daly, male sexual proprietariness, the pervasive mindset covering jealousy, entitlement to control and prevent trespass and usurpation, is a universal trait explaining why men worldwide kill women who threaten to leave and why unmarried men, who are more insecure about 'ownership' of mates, are even more likely to kill (Daly and Wilson, 1988).

Behavioural genetics, brain anatomy and neurochemistry

Whereas evolutionary psychology, with its emphasis on human universals, stresses all men's 'natural' propensity for violence against women, behavioural genetics is concerned with genetically 'caused' differences in brain and behaviour leading to the pathological factors, that create a risk of violence in the few. The research emphasis here has been on the genetics of aggression and violence, especially focussing on homicidal men, sometimes including 'lust driven' serial killers as well as uxoricidal and familicidal killers.

Research on aggression in animals and a long history of twin and adoption studies have explored the genetic/inherited proneness to violence, particularly in juveniles (Mednick *et al.*, 2003). Researchers on brain anatomy and behaviour have investigated abnormalities caused by poor development, abuse, trauma or accident and tried to identify regions in the brain linked with emotions and violent behaviour. By itself, a brain lesion is unlikely to cause violent behaviour. Indeed most people with brain damage are not violent. Dysfunctions in the brain as a result of injury or abnormality, especially to areas such as the limbic system,[1] the amygdala[2] and frontal lobes,[3] combined with, aggravated by or resulting in a chemical or hormonal underload or overload are however said to be among the factors contributing to a propensity for violence. The neurotransmitter serotonin (5-hydoxytryptamine, 5-HT) is said to act as an inhibiter to aggressive impulses, putting the brakes on violent urges fuelled by too much

[1] The limbic system is the anatomical area of the brain thought to be involved in many aspects of emotion.

[2] Placidity is thought to be linked with bilateral amygdala damage. Removal of the amygdala reduces aggression but also results in the loss of emotion. Violence is linked with abnormal electric activity in the amygdala. The rabies virus damages the amygdala and rabid animals are often aggressive.

[3] The frontal lobes, especially the orbitofrontal cortices, are thought to allow for the inhibition of aggression but damage is thought to result in aggression.

testosterone (see Chapters 14 and 15, this volume). Drugs and alcohol can have an aggravating impact as well by lowering serotonin levels in the brain thereby 'loosening the brakes' on aggression. Castrated male mice are less aggressive than non-castrated mice. Injections of testosterone into castrated mice increases aggression (Filley *et al.*, 2001). Neurochemical research on humans has similarly found links between serotonin and testosterone levels and aggression in men. Research on saliva samples taken from 692 male prison inmates in the USA found higher levels of testosterone among men committed for violent and sexual crimes than among inmates committed for other crimes (Dabbs *et al.*, 1995). Later research on 230 male prisoners accused of murder, manslaughter, robbery assault and child molestation, however, found high testosterone only seemed to be relevant to homicides (Dabbs *et al.*, 2001).

From this perspective children may inherit a proneness to violence from parents or alternatively may develop behavioural dysfunctions as a result of suffering developmental brain damage as a result of childhood abuse, although it has to be emphasised that correlations of this sort do not automatically imply causation. Nor do they indicate in the context of neurochemical and hormonal markers such as serotonin and testosterone, which way, if any, the 'chain of causation' runs. A homicidal act could potentially result in enhanced testosterone levels rather than vice versa. Sociological research suggests there are some common characteristics in the life histories of convicted killers (Dobash *et al.*, 2002; Pritchard and Stroud, 2002). Previous contact with child protection or mental health services, a history of drug or substance abuse, previous convictions for crimes of violence and child sex offending and past experience of violence and abuse are all more likely to be found among those who kill adults or children. Interestingly, men who killed their partners were less likely than other male killers to have these problems and the research into the intergenerational transmission of domestic violence is pretty inconclusive. Children living with domestic violence are often harmed. In the short term, the ill effects include fear, feelings of powerlessness, depression, suicidal and self-harming tendencies, impaired social relationships, low self-esteem, problems at school, delayed development and cognitive functions, eating disorders and aggression. An American study found that between 35% and 45% of children had symptoms of post-traumatic stress disorder (PTSD),[4] claimed to be a psychophysiological response to trauma, (but see Chapter 15, this volume). But children separated from the abusive parent on the whole recover fairly quickly from the harmful effects of living with domestic violence. The long-term effects of domestic violence on children's behaviour are unknown. Most of the research on children has either relied on what adults say about the children (rather then asking the children themselves) or it has been limited to the experiences of recently separated children living in women's refuges. Many young people who have lived with abuse or domestic violence do not grow into abusers or victims of violence as adults. Individual children respond differently. Factors influencing this include: whether the child was also physically and/or sexually assaulted by the violent parent, the individual vulnerabilities and strengths of the child, age, position in family, and the social support available outside the immediate family. A good, caring relationship with the non-violent parent has been found to be a significant factor in helping children to cope.

Feminist discussions of resilience take into account the individual strengths and vulnerabilities of children that may help them cope in the longer term but on the whole have had little to say about biological factors. Rutter has argued that psychobiological reactivity and genetic factors play a key role in determining individual differences between children's coping abilities and may explain why children in the same family situation will react differently (Rutter, 1996). It has further been argued that exposure to violence as a child could lead to neurophysiological changes and developmental brain damage that

[4] Symptoms associated with PTSD include: increased fear, hypervigilance, flashbacks and reliving the traumatic event, nightmares, sleep difficulties, depression, distress and physiological reactivity to cues of the traumatic event, problems with concentration.

increase a person's risk of becoming violent in adult life. PTSD is thought to have a long-lasting effect on neurotransmitter functions including raising levels of adrenaline, noradrenaline and glucocorticoids and lowering levels of serotonin. These neurotransmitter changes are claimed to underlie the behavioural symptoms of PTSD. In children PTSD symptoms such as high arousal and difficulty in concentrating could interfere with learning and development. Over time, it is argued, PTSD could lead to difficulties in brain development and neurophysiological traumatic stress responses and increased risk of aggression and depression. Developmental damage combined with genetic factors would further increase the risk of later aggression.

Dutton (2002) attempted to test out the argument that men who have unhappy childhoods are more likely to kill women as a result of their heightened fear of abandonment by examining serotonin and noradrenaline (known as norepinephrine in the USA) levels in convicted wife-killers in Canada. Most of the men were said to have killed their partners in response to her threat to leave. Dutton argued that attachment dysfunction at ages 1.5 to 2 years could interfere with the development of the orbitofrontal cortex leading to the low levels of serotonin and high levels of noradrenaline found in the convicted wife-killers, thus linking early trauma with a rage response to abandonment and homicide.

In summary it can be said that evolutionary psychologists and behavioural geneticists in particular offer a range of explanations as to how violent or homicidal behaviour may be linked to biology. Some research points towards external forces, brain damage resulting from injury, birth defect or poor attachment in parenting as resulting in lasting neurobiological changes that in turn contribute to the terror, fear, rage and violence in response to the threat of abandonment. Whether the result of genetic or developmental factors, such experiences create a 'short fuse', lowering a person's tolerance to stresses, provocations or violent impulses. Biological factors are thus viewed as contributing to, rather than determining, the *risk* of aggression and violence.

A 'fresh' approach?

The claim that men have a natural proneness to rape and abuse women, even if proven, would not absolve them of responsibility for doing so (see the chapters in this volume by Stephen Sedley (Chapter 7) and Sandy McCall Smith (Chapter 6) on issues of legal responsibility). But although evolutionary psychologists such as Thornhill and Palmer argue that this means that society has to try harder to stop the violence, taking a 'fresh' approach tackling the social and biological factors together, their claims imply an inevitability about violence, some or all men being prone to behave badly unless they are controlled. Violence and anti-social behaviour have become important political issues in the context of the rolling back and modernisation of state welfare services. In this context of changing governance the trend to perceive crime, violence, poverty, family breakdown and unemployment as stemming from behavioural deficiencies amongst the poor and socially excluded has brought a further individualisation of responsibility for crime and helped (re)open the door to 'biological fatalism'. Worries that such apparently scientifically based arguments might be used to justify the apparent inevitability of inequalities based on class, race and gender are hard to cast aside, especially if the claims are hard to replicate (Fausto-Sterling, 2001). Extrapolating studies on the aggression, sexual behaviour and courtship rituals of dung beetles, woodcocks, nursery web spiders (as in Thornhill and Palmer, 2000b) and mice (Filley *et al.*, 2001) to apparently 'explain' human males' domestic abuse of women seems tenuous at best. Research on human males in captive experimental groups (in prisons), is mostly retrospective, coming well after the violent events. Studies of homicide usually take a one-sided view of events as the victims are no longer there to say their piece.

Objections could also be sustained in the face of the universalising theories of femicide offered, especially by evolutionary psychologists. Homicide rates vary considerably from place to place and from time to time. Cross-national comparisons are fraught with difficulties due to variations in definitions, record-keeping and criminal justice responses. Evolutionary psychologists acknowledge these methodological difficulties but nonetheless explain the variations in marital femicide rates as resulting from the levels of sexual insecurity males feel/perceive within different societies or cultures. Some societies and cultures, it is argued, create contexts where men feel more insecure about their sexual 'ownership' of women. For example, evolutionary psychologists assert that high levels of divorce and relationship breakdown would cause men to feel insecure and more prone to use violence to control their partners. Also, where men in a population outnumber the eligible, reproductive-age females, male competition for females increases and similarly, it is assumed, there will be higher levels of violence.

This biologically reductionist account cannot convincingly explain the very big variations that exist in marital femicide rates. The marital femicide rates in the USA are between five and ten times higher than the rates in parts of Europe, yet the cultural or socio-legal factors that might pose a threat to male sexual proprietorial security, such as rates of divorce and partner separation, are fairly similar. Unpicking the data on child homicides further suggests the limited relevance of biology, genetics or evolutionary fitness in explaining why children are killed more in some countries than in others. Rates of child homicide are declining in all industrialised countries apart from France and the USA. Once top of the child homicide league (World Health Organisation data), England has seen a decline in child homicide rates since 1974 and now has the third lowest rate. This has happened while rates of marital femicide have not changed and children's poverty and their experiences of living in female-headed single-parent families has dramatically grown. Even where there has been no (apparent) prior history of domestic violence, evolutionary psychologists such as Wilson and Daly rely on male sexual proprietariness as an explanation. Half of all Canadian men who kill their wives and children also kill themselves. These 'male despondent killers' according to Wilson and Daly usually do not have a history of prior violence to partners but they still act in a male protective manner, killing themselves and their families as the only solution to their problems (Wilson and Daly, 1999). All problems that may (or may not) 'drive' men to kill women and children thus stem ultimately from male sexual proprietariness. Some femicides do not fit the theory too well. Lots of men rape and abuse their sisters. A study of woman killing in Jordan in 1995 found that most domestic violence homicides of women were 'honour crimes' committed by brothers on sisters (Kulwicki, 2002). Men commonly assault and kill their partners during pregnancy. Sexuality may well play a part but conceptual contortions would be needed to explain this as all linked to evolutionary fitness.

IMPLICATIONS FOR GOVERNANCE

The theoretical and empirical foundations of the new biology of crime are shaky but the possible impact upon policy and practice warrants some discussion. The new biologism could be linked, in a post-Foucauldian nightmare vision, to the disciplinary expansion of 'bio-power' within a public health regime that combines the identification of violence-prone individuals with their treatment and control. Under this regime, adults, children and even the unborn could be screened for risk of a violent predisposition. Those identified as being at risk of or having committed acts of violence could be 'helped', medically or therapeutically treated or re-educated out of violent behaviour. This could have a political appeal to welfare minimalists wanting to make savings for the public purse because resources could be targeted towards those sectors of the population identified as being most 'at risk' (although treatment tends to be more costly than incarceration). The range of treatment options would expand to include the possibility of diet and chemical treatment alongside imprisonment, offender re-education or therapy. Already, in Germany sex offenders can undergo 'corrective' castration and in the USA sexually motivated murderers are given injections of female hormones to help quell the effects of excess testosterone.

It would be easy to run amok with this vision of the biologically screened and risk-managed future. Testing and risk assessment, especially lethality assessments, are already important in North American policy on domestic and sexual violence. In England, risk assessment and treatment for violent men has been gaining ground. The argument, bolstered by biological claims, that the truly dangerous need to be both contained and treated, can be seen creeping in to recent legislative efforts. The 2003 sex offenders' bill for instance, proposed indeterminate sentences for sex offenders with release not possible without a risk assessment. The public health message about men's violence towards women has however been promoted by feminists as well as by the new biologists and the implications for governance need to be considered.

The preventive screening of individuals for a biological or inherited propensity for violent behaviour could aggravate an already observed tendency in the trend from government to governance, to shift responsibility for crime management and crime control from elected government and national agencies towards the local and on to individuals. It is unlikely that screening convicted or non-convicted men for femicidal proneness would help eliminate the domestic abuse of women. Biologically informed screening techniques may be used to identify further risk in the very few men imprisoned for abusing their partners or children. The new biology of violence might be applied to monitor or even treat or manage the behaviour of previously convicted partner abusers but this would not contribute much towards the practical work of probation officers nor to the resources needed by them to do their jobs. Screening might become more commonplace as a result of risk-averse behaviour and the fear of litigation. Agencies fearing litigation such as social work, holiday camps, hospitals, schools, hostels, probation services and possibly church groups would be more likely to take up screening opportunities and risk assessments than individual women hoping for a pre-nuptial vetting of their partners. The most femicidally prone men, the 'ordinary' men who live in the house, would be those least likely to be subjected to, or to consent to, any risk assessment or screening procedure. Screening for a lethal potentiality towards violence is much more likely to lead to the singling out of some easily identified violent men – the dangerous strangers and paedophiles – while men's everyday violence to their partners continues unnoticed.

Biologically determinist arguments paint an image of men with 'short fuses' driven uncontrollably by either lust or a rage provoked by a woman's behaviour. This image matches what defendants, and their counsel, often argue in court. Thomas Corlett for example was reported to have strangled his wife because she moved the mustard pot. When Corlett was tried for murder a psychologist argued in court on his behalf that he was suffering from diminished responsibility when he killed his wife, having been driven into a rage by his wife's disruption of his obsessionally house-proud rituals (Lees, 1997: 165). The image does not entirely match what survivors say about domestic violence and the part that control plays in it. Many survivors say they tried to manage the abuser's behaviour and keep the peace, sometimes to such an extent that they found themselves 'walking on eggshells', constantly watching and trying to avoid the abuse. Indeed Walker (1984) suggests that, although an understandable response, over time a woman's efforts to keep the peace by adapting her behaviour to appease the abuser can make the violence worse. Although very few violent men do end up killing their partners, suggestions that they only do so when 'out of control' or provoked to be femicidal should be treated warily. Focussing on the man's short fuse could dangerously shift responsibility for managing his behaviour back on to his victim. She would have to manage her behaviour in order to manage him. This would be a great backward step in preventing violence as it would shift the focus away from helping women to get the resources they need to be free of violence.

How the new neurogenetic approaches might be used in practice to control an identified risk of violence against women raises concerns. There are in reality few treatment options available for violent men. Surgical treatment would be unsuitable for mass programmes.

Violent men would be unlikely to consent in great numbers and there would be human rights objections against any mandatory treatment. Chemical castration by injection of so-called 'female' hormones would be more feasible and less expensive but is based on the doubtful assumption that such injections will reduce or counterbalance the levels of 'male' hormones supposed to promote violence and hence reduce risk. This might be an option chosen by 'dangerous' offenders to boost their risk assessment scores sufficiently to gain a release from an otherwise indefinite sentence. But as few wife- or partner-killers are likely to be so sentenced, few would need to put themselves forward for this 'treatment'. Other treatment options include cognitive therapy and re-education into different patterns of behaviour. These are areas where expansion of the public health regime would be most likely. There is some research evidence that suggests that projects that aim to re-educate violent men out of abusive behaviour are more effective in stopping violence than traditional criminal justice sanctions alone but the longer-term effects are not known. There are no studies showing whether or not non-violent behaviour is sustained over time. Supporters of re-educative work with violent men note the dangers that faith in a 'cure' may pose to victims. Re-educative work is where the new biological approaches are most worrying. Thornhill and Palmer's 'fresh approach' to rape education for instance means applying an evolutionary psychologists' perspective to the understanding of rape so that women and girls are told about why men rape. Thus having explained the theory of sexual fitness group leaders are advised to warn women and girls that, as rape is motivated more by lust than by violence, they should avoid dressing 'provocatively'. The total irrelevance of this advice to women and girls of all ages, including the children who are sexually abused, was long ago parodied by the London Rape Crisis Centre's advice for women on what to wear – no clothes and all clothes 'provoke' some men to rape.

Some thoughts need to be given to the implications that an anti-violence public health programme might raise for children and young people. In practice both feminists and behavioural geneticists put forward remarkably similar proposals for dealing with the impact of domestic violence on children. Both favour wider monitoring and the early identification of risk by healthcare workers combined with a programme of preventive work with children in schools. However, approaches to risk assessment based on the claims of behavioural geneticists would tend to support targeting resources towards those in the population thought likely to be most pathological and violence-prone – most likely the 'underclass' and socially excluded – as opposed to seeing it as part of a continuum. Healthy parenting policies are presently focussed predominantly on the poor. History shows the need to be cautious of the net-widening consequences of watching the young for potential criminality. The result might be a labelling of more disadvantaged children as problem children who are violence-prone, while violence in middle-class families persists unchecked.

By contrast feminists advocate screening and risk assessment only if it is part of a broader programme of social change and public education where violence would become unacceptable. Screening policies are geared less to lethality and more to enhancing safety and ensuring that an individual's needs are taken into account. Expert assessments are less important than the adult or child's account of their experiences and needs. Whilst not denying that children in the context of abuse may be less able to talk about their own needs and fears, feminist perspectives have stressed an approach that aims to maximise the child's involvement in decision-making. There is scope here to direct attention away from biologism's expert driven emphasis on pathology and lethality towards the assessment of needs, risks and resources. There are many reasons to be wary of the claims of evolutionary psychologists and behavioural geneticists on femicide and violence. However, they cannot simply be dismissed and ignored as deterministic positivism. In particular it is important that feminism continues to challenge the pathologisation of violence and any attempts to argue that it is biologically determined. A starting point might be to redirect research on violence against women, shifting attention away from the solely femicidal men imprisoned for killing women, to explore the sexual violence continuum. This would require looking at 'normal' masculinities and at the high proportions of 'normal' men who abuse women and girls, and questioning in a more reflexive way the relationship between 'the natural' and 'the pathological'.

ACKNOWLEDGEMENT

I dedicate this chapter to Sue Lees who, sad to say, died in September 2002. Sue was a tireless researcher, educator, campaigner and activist against violence against women. Her work on femicide and on rape was an inspiration to me and to many others. Sue encouraged me to consider the issues addressed in this piece.

9 Genes, responsibility and the law

PATRICK BATESON

It is often supposed that all shreds of human agency succumb in the face of advances in the understanding of evolutionary process, genetics and brain function. Conventional wisdom collapses and all responsibility for the consequences of our actions is diminished to the point at which, it is claimed, no blame can be attached to anything we do.

Or so the argument goes, but is it really the case that science has had such serious implications for the way we should think about our own capacity for choice? The importance of the emotions in controlling human behaviour certainly suggests to some that all of us are in the grip of our instincts and our genes. We seem to be surrounded by examples of irrational behaviour, such as when people are in love, in lynching mode or maddened with war fever. The brain (and the genes that contribute to its construction) are such that, when people make conscious choices, they don't *really* know what they are doing and if so the presumptions of law, morality and common sense must be wrong.

In 1979 the Mayor of San Francisco and one of his officials were gunned down by one Dan White. At his trial White was convicted of manslaughter instead of the first-degree murder of which he was accused. His lawyers produced an original argument which came to be known as the 'Twinkie defence'. Dan White was addicted to a sugary junk food known as Hostess Twinkies and his lawyers managed to persuade the court that his brain had been so deranged by them that he should not be held fully responsible for his actions. The worrying thing about this argument is not that the scientific basis for it is so

The New Brain Sciences: Perils and Prospects, ed. D. Rees and S. Rose.
Published by Cambridge University Press.

slender. It is part of a pattern which steadily erodes the conception of humans having intentions for which they can be held responsible.

Nevertheless, good examples of diminished responsibility pose the problem of how they are to be separated from the cases that are much more equivocal. It is, alas, all too obvious that humans do stupid things that run counter to their own best interests and the interests of those around them. It is obvious too that some forms of behaviour that may have benefited humans in the past have become dysfunctional in the radically different modern world. Behaviour such as seeking out and receiving pleasure from eating sweet or fatty foods may well have been vital in a subsistence environment, but in a well-fed society it does more harm than good. Part of the problem could be that human behaviour was adapted to circumstances in which people no longer live or to those which are rendered non-functional by conditions in which they now find themselves.

Gambling, which often ruins lives, seems wholly irrational but could perhaps have made sense in a world in which the delivery of rewards is rarely random. If you have done something that produced a win, it is usually highly beneficial to repeat what you did – except when you get into a casino. Similarly, the tendency of parents to protect their children from all contact with unknown people after hearing of a child murder on television would have been beneficial in a small community where such news might represent real danger. In the modern context, such risk-averse behaviour in a society in which the incidence of child murder has remained constant for decades merely impoverishes their children's development. Even though they were once doubtless well-designed for the world in which humans evolved, the emotions may now play havoc with our lives. They have become what Stuart Sutherland called 'The Enemy Within' (Sutherland, 1992).

The importance of the emotions in the organisation of behaviour raises big questions about brain function, the relevance of evolutionary biology in understanding behaviour, and the role of genes in the relatively invariant pattern of individual development. I have attempted to deal with them at some length in the book I wrote with Paul Martin,

Design for a Life (Bateson and Martin, 2000) and will only touch on the main issues here.

SELFISH GENES

Are humans blundering robots, programmed to do all those things that keep them alive and help them to pass on their genes to another generation? Famously, Richard Dawkins suggested that the way to understand evolution is not in terms of the needs of the individual (or the group or the species), but in terms of the 'needs' of genes (Dawkins, 1976). Genes recombine in each generation to form temporary federations. The alliance forms an individual organism. By reproducing, individuals serve to perpetuate the genes which in the next generation recombine in some other kind of alliance. Genes can thus be conceived of as 'intent' on replicating themselves by the best possible means.

The gene's eye view grew up because it made sense of those cases in which the consequences of an act favoured the survival of genetically related individuals rather than the actor. Dawkins was clearly and deliberately using a device to aid thought about such evolutionary dynamics when he attributed motives to genes. He did not suppose that genes really have intentions. His writings made their impact, I suspect, because most people more readily get their minds round the workings of a complex system when it is personified. They imagine how that system strives to reach a specific end-state. However, the mental device has its shortcomings when the parable is interpreted too literally. Attributing selfishness to genes neither implies that they have intentions, nor does it give us any insight into the mechanisms of evolution.

The ease with which those who have been influenced by Dawkins move from the language of goals to the language of causality explains why many biologists have become nervous about the idea of genes being treated as the units of selection in Darwinian terms. Darwin had used his metaphor of 'natural selection' because he had been impressed by the ways in which plant and animal breeders selected the characters that they sought to perpetuate. He did not suggest that biological units evolved as a result of the intentions of those units. Nor did he wish to imply that the analogy with artificial selection of plant and animal breeders was an external hand-picking for breeding some forms and discarding others. To understand the evolutionary process fully we must know what generates the variation in the raw material for differential survival and reproductive success. We must also understand the necessary conditions for re-creating the characteristics of the successful entity in the next generation. The relationships will commonly be defined by genes but also by a great many non-genetic factors.

WHERE DOES BEHAVIOUR COME FROM?

With increasing frequency the media report the discovery of a gene 'for' some distinct human characteristic, such as learning foreign languages, athletic prowess or male promiscuity. Yet it is obvious that experience, education and culture make a big difference to how people behave, whatever their genetic inheritance. Why is it that behavioural and psychological development is so often explained in terms of the exclusive importance of one set of factors, either genetic or environmental? The answer can lie partly in the character of scientific debates. If Dr Jones has overstated her case, then Professor Smith might feel bound to redress the balance by overstating the counter-argument. The confusions are amplified because of the way in which scientists analyse developmental processes. When somebody has conducted a clever experiment demonstrating an important long-term influence on behaviour, they have good reason to feel pleased. It is easy to forget about all those other influences that they had contrived to keep constant or which play no systematic role. Consequently, debates about behavioural and psychological development often degenerate into sweeping assertions about the overriding importance of genes (standing in for 'nature') or the crucial significance of the environment (which then becomes 'nurture'). Understanding behavioural development means understanding the biological and psychological processes that build a unique adult from a fertilised egg. It does not mean trying to explain human behaviour in terms of the conventional opposition between nature (genes) and nurture (environment).

No serious biologist can doubt that genes matter. Plant and animal breeders know well that many of the characteristics that matter to them are inherited, in the sense that a new set of progeny will resemble individuals in the ancestral pedigree of that plant or animal more than they resemble progeny from some other pedigree. Long before genes and DNA were discovered, breeders took this as a bountiful fact of life, even though they had no idea how inheritance worked. For centuries, and in some cases millennia, domestic animals have been artificially selected by humans for breeding because they exhibit specific physical or behavioural features that are regarded as desirable. Dogs, in particular, have for many centuries been bred for their behavioural characteristics. The sheepdog is especially sensitive to the commands of humans, waiting until the shepherd gives it a signal to start herding the sheep. Another breed, the pointer, behaves in a way that helps in sports shooting. When the pointer detects the smell of a game species such as grouse, the dog stops in its tracks, stiffly orientated towards the bird. Valued behavioural characteristics such as these are clearly inherited, do not need to be taught (at least, not in their most basic form) and are quickly lost if breeds are crossed with others. Humans may also reveal through their children how particular characteristics are inherited. Two healthy parents from a part of the world where malaria is rife may have a child who develops severe anaemia. Both parents carry a gene that does have some effect on red blood cells, protecting them against the malarial parasite which enters red blood cells for part of its life cycle. However, a double dose of this recessive gene leads to the red blood cells collapsing from their normal biconcave disc shape into strange sickle-like shapes. The child who receives this genetic legacy has sickle-cell anaemia.

Evidence for genetic influences on human behaviour is usually indirect. It is bound to be so because naturally occurring breeding experiments are rare, and deliberate breeding experiments in the interests of genetic research would obviously be intolerable in most societies. What is more, many genes are involved in the great majority of family likenesses, whether physical or behavioural.

Genes matter, but so does experience. The most cursory glance at humanity reveals the enormous importance of each person's experience, upbringing and culture. Look at the astonishing variation among humans in language, dietary habits, marriage customs, child-care practices, clothing, religion, architecture, art and much else besides. Nobody could seriously doubt the remarkable human capacity for learning from personal experience and learning from others. Sometimes the size, metabolism and behaviour of an individual may be adapted to the conditions prevailing early in his or her life. So the person may have a 'thrifty phenotype' if the mother's nutritional state during pregnancy is poor and an 'affluent phenotype' if the mother was well-fed and in good health. The trajectory of development is triggered in this case by cues provided by the mother about the state of the environment.

Sometimes early experience can have effects which were not anticipated. In the 1960s, great efforts were made in the USA to help people living in difficult and impoverished conditions. A large government programme known as Headstart was designed to boost children's intelligence by giving them educational experience before starting school. In the event, the Headstart programme did not seem to have the substantial and much hoped-for effects on intelligence, as measured by intelligence quotient (IQ). Children who had received the Headstart experience displayed an initial, modest boost in their IQ scores, but these differences soon evaporated after a few years. The fashionable response was to disparage such well-meaning efforts to help the disadvantaged young.

Later research, however, has revealed that some of the other effects of the Headstart experience were long-lasting and of great social significance – greater, in fact, than boosting IQ scores. Several long-term follow-up studies of children who had received pre-school training under Headstart found that they were distinctive in a variety of ways, perhaps the most important being that these individuals were much more community-minded and less likely to enter a life of crime (Schweinhart *et al.*, 1993). Headstart produced lasting benefits for the recipients and society more generally, but not by raising raw IQ scores. Evidence for the long-term benefits of early educational intervention has continued to accumulate (Yoshikawa, 1995). Studies like these raise many questions about the ways in which early experiences exert their effects, but they do at least show how important such experiences can be.

THE CHARACTER OF THE INTERPLAY

The importance of both genes and environment to the development of all animals, including humans, is obvious. This is true even for apparently simple physical characteristics, let alone complex psychological variables. Take myopia (or short-sightedness) for example. Myopia runs in families, suggesting that it is inherited. But it is also affected by the individual's experience. Both a parental history of myopia and, to a lesser extent, the experience of spending prolonged periods studying close-up objects increases the probability of a child becoming short-sighted.

A more interesting case is musical ability, about which strong and contradictory views are held. Popular beliefs about the origin of special talents are generally that they are inherited. Dissociation between general intellectual capability and musical ability is strongly suggested by the phenomenon of the musical idiot savant – an individual with low intelligence but a single, outstanding talent for music. Such people are usually male and often autistic. Their unusual gift – whether it be for music, drawing or mental arithmetic – becomes apparent at an early age and is seldom improved by practice. However, the main factors fostering the development of musical ability form a predictable cast: a family background of music; practice (the more the better); practical and emotional support from parents and other adults; and a good relationship with the first music teachers.

Practice is especially important, and attainment is strongly correlated with effort. A rewarding encounter with an inspirational teacher may lock the child in to years of effort, while conversely an unpleasant early experience may cause the child to reject music, perhaps for ever. Here, as elsewhere, chance plays a role in shaping the individual's development. Like many other complex skills, musical ability develops over a prolonged period; and the developmental process does not suddenly stop at the end of childhood. Expert pianists manage to maintain their high levels of musical skill into old age despite the general decline in their other faculties. They achieve this through copious practice throughout their adult life; the greater the amount of practice, the smaller the age-related decline in musical skill. Practice not only makes perfect, it maintains perfect.

Is it possible to calculate the relative contributions of genes and environment to the development of behaviour patterns or psychological characteristics such as musical ability? Given the passion with which clever people have argued over the years that either the genes or the environment are of crucial importance in development, it is not altogether surprising that the outcome of the nature–nurture dispute has tended to look like an insipid compromise between the two extreme positions. Instead of asking whether behaviour is caused by genes or caused by the environment, the question instead became: 'How much is due to each?' In a more refined form, the question is posed thus: 'How much of the variation between individuals in a given character is due to differences in their genes, and how much is due to differences in their environments?'

The nature–nurture controversy appeared at one time to have been resolved by a neat solution to this question about where behaviour comes from. The suggested solution was provided by a measure called heritability. The meaning of heritability is best illustrated with an uncontroversial characteristic such as height, which is clearly influenced by both the individual's family background (genetic influences) and nutrition (environmental influences). The variation between individuals in height that is attributable to variation in their genes may be expressed as a proportion of the total variation within the population sampled. This index is known as the heritability ratio. The higher the figure, which can vary between 0 and 1.0, the greater the contribution of genetic variation to individual variation in that characteristic. So, if people differed in height solely because they differed in their genes, the heritability of height would be 1.0; if, on the other hand, variation in height arose entirely from individual differences in environmental factors such as nutrition then the heritability would be 0. The relative influences may be estimated by comparing identical (same genes) and non-identical (different genes) twins reared together (and therefore presumably in the same environment), and sometimes also identical twins separated for adoption in different families (same genes, different environments). More than thirty twin studies, involving a total of more than 10 000 pairs of twins, have collectively produced an estimated heritability for IQ of about 0.5 (ranging between 0.3 and 0.7). Twin and adoption studies of personality measures, such as sociability/shyness, emotionality and activity level, have typically produced heritabilities in the range 0.2 to 0.5 (Plomin *et al.*, 1997).

Calculating a single number to describe the relative contribution of genes and environment has obvious attractions. Estimates of heritability are of undoubted value to animal breeders, for example. Given a standard set of environmental conditions, the genetic strain to which a pig belongs will predict its adult body size better than other variables such as the number of piglets in a sow's litter. If the animal in question is a cow and the breeder is interested in maximising its milk yield, then knowing that milk yield is highly heritable in a particular strain of cows under standard rearing conditions is important.

But behind the deceptively plausible ratios lurk some fundamental problems. For a start, the heritability of any given characteristic is not a fixed and absolute quantity – tempted though many scientists have been to believe it is. Its value depends on a number of variable factors, such as the particular population of individuals that has been sampled. For instance, if heights are measured only among people from affluent backgrounds, then the total variation in height will be much smaller than if the sample also includes people who are small because they have been undernourished. The heritability of height will consequently be larger in a population of exclusively well-nourished people than it would be among people drawn from a wider range of environments. Conversely, if the heritability of height is based on a population with relatively similar genes – say, native Icelanders – then the figure will be lower than if the population is genetically more heterogeneous; for example, if it includes both Icelanders and African pygmies. Thus, attempts to measure the relative contributions of genes and environment to a particular characteristic are highly dependent on who is measured and in what conditions.

Another problem with heritability is that it says nothing about the ways in which genes and environment contribute to the biological and psychological processes of development. This point becomes obvious when considering the heritability of a characteristic such as 'walking on two legs'. Humans walk on fewer than two legs only as a result of environmental influences such as war wounds, car accidents, disease or exposure to teratogenic toxins before birth. In other words, all the variation within the human population results from environmental influences, and consequently the heritability of 'walking on two legs' is zero. And yet walking on two legs is clearly a fundamental property of being human, and is one of the more obvious biological differences between humans and other great apes such as chimpanzees or gorillas. It obviously depends heavily on genes, despite having a heritability of zero. A low heritability clearly does not mean that development is unaffected by genes.

If a population of individuals is sampled and the results show that one behaviour pattern has a higher heritability than another, this merely indicates that the two behaviour patterns have developed in different ways. It does not mean that genes play a more important role in the development of behaviour with the higher heritability.

Yet another serious weakness with heritability estimates is that they rest on the spurious assumption that genetic and environmental influences are totally independent. In many cases this assumption is clearly wrong. For example, in one study of rats the animals' genetic background and their rearing conditions were both varied; rats from two genetically inbred strains were each reared in one of three environments, differing in their richness and complexity. The rats' ability to find their way through a maze was measured later in their lives. Rats from both genetic strains performed equally badly in the maze if they had been reared in a poor environment (a bare cage) and equally well if they had been reared in a rich environment filled with toys and objects. Taken by themselves, these results implied that the environmental factor (rearing conditions) was the only one that mattered. But it was not that simple. In the third type of environment, where the rearing conditions were intermediate in complexity, rats from the two strains differed markedly in their ability to navigate the maze. These genetic differences only manifested themselves behaviourally in this sort of environment. Varying both the genetic background and the environment revealed an interplay between the two influences.

An overall estimate of heritability has no meaning in a case such as this, because the effects of the genes and the environment do not simply add together to produce the combined result. The effects of a particular set of genes depend critically on the environment in which they are expressed, while the effects of a particular sort of environment depend on the individual's genes. Even in animal breeding programmes that use heritability estimates to practical advantage, care is still needed. If breeders wish to export a particular genetic strain of cows which yield a lot of milk, they would be wise to check that the strain will continue to give high milk yields under the different environmental conditions of another country. Many cases are known where a strain that performs well on a particular measure in one environment does poorly in another, while a different strain performs better in the second environment than in the first.

Any scientific investigation of the origins of human behavioural differences eventually arrives at a conclusion that most non-scientists would probably have reached after only a few seconds' thought. Genes and the environment both matter. The more subtle question about how much each of them matters defies an easy answer; no simple formula can solve that conundrum. The problem needs to be tackled differently.

DEVELOPMENTAL PROCESSES

The idea that genes might be likened to the blueprint of a building must be scrapped immediately – at least as far as behaviour is concerned. The idea is hopelessly misleading because the correspondences between plan and product are not to be found. In a blueprint, the mapping works both ways. Starting from a finished house, the room can be found on the blueprint, just as the room's position is determined by the blueprint. This straightforward mapping is not true for genes and behaviour, in either direction. The language of a gene 'for' a particular behaviour pattern so often used by scientists, is exceedingly muddling to the non-scientist (and, if the truth be told, to many scientists as well). What the scientists mean (or should mean) is that a genetic difference between two groups is associated with a difference in behaviour. They know perfectly well that other things are important and that, even in constant environmental conditions, the developmental outcome depends on the whole gene 'team'. Particular combinations of genes have particular effects, in much the same way as a particular collection of ingredients may be used in cooking a particular dish; a gene that fits into one combination may not fit into another. Unfortunately, the language of genes 'for' characters has a way of seducing the scientists themselves into believing their own soundbites.

The adult human brain has around one hundred thousand million (10^{11}) neurons, each with hundreds or thousands of connections to other neurons. A diagram of even a tiny part of the brain's connections would look like an enormously complex version of a map of the London Underground railway system. The brain is organised into subsystems, many of which are dedicated to different functions which are run separately but are integrated with each other. Since the behaviour of the whole animal is dependent on the whole brain, it will be obvious why it is not sensible to ascribe a single aspect of behaviour to a single neuron, let alone a single gene. The pathways running from genes to neurons and thence to behaviour are long, full of detours, with many other paths joining them and many leading away from them.

Nothing happens in isolation. The products of genes and the activities of neurons are all embedded in elaborate networks. Each behaviour pattern or psychological characteristic is affected by many different genes, each of which contributes to the variation between individuals. In an analogous way, many different design features of a motor car contribute to a particular characteristic such as its maximum speed. Conversely, each gene influences many different behaviour patterns. To use the car analogy again, a particular component such as the system for delivering fuel to the cylinders may affect many different aspects of the car's performance, such as its top speed, acceleration and fuel consumption. The effect of any one gene also depends on the actions of many other genes.

Development is not like a fixed musical score which specifies exactly how the performance starts, proceeds and ends. It is more like a form of jazz in which musicians improvise and elaborate their musical ideas, building on what the others have just done. As new themes emerge, the performance acquires a life of its own, and may end up in a place none could have anticipated at the outset. Yet it emerges from within a system of rules and the constraints imposed by the musical instruments.

The commonly used image of a genetic blueprint for the development of behaviour is misleading because it is too static, too suggestive that adult organisms are merely expanded versions of the fertilised egg. In reality, individuals play an active role in their own development. Even when a particular gene or a particular experience is known to have a powerful effect on the development of behaviour, biology has an uncanny way of finding alternative routes. If the normal developmental pathway to a particular form of adult behaviour is impassable, another way may often be found. The individual may be able, through its behaviour, to match its environment to suit its own characteristics – a process dubbed 'niche-picking'. At the same time, an activity such as play, a feature of development shared by the young of many species, increases the range of available choices and, at its most creative, enables the individual to control the environment in ways that would otherwise not be possible.

CONCLUSION

Whatever claims are made for the role of the emotions in both guiding and misguiding behaviour, humans have shown remarkable capacity for rapid change. The transformation of man-made environments, and subsequent human adaptations to them, have been abrupt and recent, relative to human evolutionary history. The earliest forms of civilisation, in the shape of systematic farming, emerged less than 10 000 years ago. The first written records appeared 6000 years ago in Mesopotamia and China, and the wheel was invented not long afterwards. Industrialised societies started to emerge within the past 200 years and computers only became ubiquitous in the later part of the twentieth century. The changes in environmental conditions from those in which humans evolved have been radical and have occurred when genetic change has been negligible.

Opportunism plays an important role in driving historical change. Humans are perfectly capable of appreciating the value of their own experiments, and the emerging effects have had an extraordinary influence on human history. The combination of spoken language, which has obvious utility in its own right, and manual dexterity in fashioning tools, which also has obvious utility, occurred at a particular and relatively recent moment in evolutionary history to generate written language. The discovery of written language took place several times and in several forms in different parts of the world, with ideas represented by pictures or spoken sounds represented by symbols. Once invented, the techniques were quickly copied and became crucial elements of modern civilisation. It was that active combining of different capacities that started the whole remarkable cultural sequence of events (see Chapter 2, this volume, which analyses these developments in considerable detail).

Such opportunism is directly relevant to the matter in hand, the discussion of decision, freedom of will and moral responsibility. In general, people respond appropriately to the consequences of their own behaviour. If they are punished for what they did, they are much less likely to repeat those actions and if they are rewarded, they are likely to repeat them. If an understanding of the likely consequences can be achieved without actually performing the act, then a person who knows that they will be punished or rewarded for certain acts is bound to be influenced by such assessments. People are able to make free choices because their choices are enormously beneficial to their welfare and ultimately to their survival. Of course, people may be surprised by the consequences of their own actions. Nevertheless, planning before doing is of great advantage to the individual.

When science meets the law, those who appeal to biology like to suggest that all sorts of things may lead to diminished responsibility – evolutionary history, faulty genes, faulty development, environmentally induced malfunctioning of the brain through disease, environmental accident or exposure to drugs. Clearly, the emotions can be derailed by a host of factors and they can all provide handy defences for lawyers. Their use, however, requires agreement on what should be the default position when somebody behaves in socially damaging ways. My own view is that we should assume intentionality, and hence responsibility until we have very good reason to think otherwise.

Part IV Stewardship of the new brain sciences

10 The neurosciences: the danger that we will think that we have understood it all

YADIN DUDAI

In 1953, the same year in which he had operated on the brain of the famous amnesic H. M., the American neurosurgeon W. B. Scoville described the major achievements of contemporary neurosurgery, while at the same time disclosing his aspirations for the future:

> We have isolated, by the 'undercutting' technique, the anterior cingulate gyrus and the posterior orbital cortex in a series of fractional lobotomies performed on schizophrenic and neurotic patients. More recently, we have both stimulated and resected bilaterally various portions of the rhinencephalon in carrying out medial temporal lobectomies on schizophrenic patients and certain epileptic patients . . . orbital isolation has given a most gratifying improvement in depression, psychoneuroses, and tension states . . . Who knows but that in future years neurosurgeons may apply direct selective shock therapy to the hypothalamus, thereby relegating psychoanalysis to that scientific limbo where perhaps it belongs? And who knows if neurosurgeons may even carry out selective rhinencephalic ablations in order to raise the threshold for all convulsions, and thus dispense with pharmaceutical anticonvulsants? (Scoville, 1954).

One doesn't need to be a neuroanatomist, with a detailed understanding of these brain structures to appreciate the optimistic tone. These were, no doubt, the high days of psychosurgery. They began in

The New Brain Sciences: Perils and Prospects, ed. D. Rees and S. Rose.

the 1930s when the Portuguese neurologist Egas Moniz attempted to treat mental illness by severing neural tracts in the frontal cortex. The approach became astonishingly widespread, apparently not without support from the popular press. Moniz was even awarded the Nobel Prize in 1949 for developing it. Altogether, between the 1940s and the 1960s, many thousands of patients were lobotomised with little hesitation, mainly in the Americas and Europe but in other continents as well, crippling forever even further the mentally ill, but sometimes even the mentally not so ill, at the discretion of physicians, and sometimes also with the encouragement of ignorant bureaucrats.

This type of treatment is now passé, partly due to the disappointing outcome, partly due to the introduction of less-or-more effective psychopharmacological agents. What was a standard procedure for many psychiatric disorders is now considered unacceptable and furthermore outlawed in many countries. In others it is still reserved for very special cases. In some countries patients who underwent brain lobotomy are entitled to compensation by the state. We do seem to appreciate now how much we did not know only fifty years ago. Returning to Scoville's predicition, cutting the connections to the brain's temporal lobe is still indicated as the last resort in drug-refractory, catastrophic epilepsy, including for young children – not without occasional criticism. Interestingly, even the most famous of such operations, on a patient known as H. M. which is discussed in every neuroscience textbook, significantly relieved the epilepsy but left the patient deeply amnesic for ever, was admitted to be a 'frankly experimental operation' (Scoville and Milner, 1957); no localised epileptic foci to be removed were identified in his brain by pre-operative studies.

Lobotomy is mostly gone, yet the attitude that has once culminated in lobotomy is not. It is concealed in recent discussions on the present and future of brain research. Only today the surgeon's knife has been replaced by ideas that involve genetic manipulations, brain transplantations, even neurosilicon hybrids. The name of the game is the same: irreversible intervention in the human brain, assuming an understanding of how the brain works and what the intervention will do to physiology, emotion and cognition. The bare truth is that this assumption is unfounded. This is in spite of the remarkable achievements of the brain sciences, and in spite of some hypes of modern biology, reinforced and augmented by commercial interests on the one hand, news-hungry press on the other, and occasional personal hubris of overenergetic scientists in between. In this chapter I will briefly illustrate what is meant by the 'lobotomy attitude' in contemporary neuroscience. My very limited selection of examples involves state-of-the-art methodologies that also play a major role in forecasts of developments in medicine and society.

NEUROGENETIC INTERVENTION

Though few will doubt that gene therapy should be extensively explored for its potential to prevent or alleviate genetic defects and neurodegenerative diseases, its risks should not be underestimated. Even apart from the social concerns that many have expressed, pure scientific considerations expose the problem. This should be particularly taken into account in discussions that predict and promote the use of genetic manipulations to augment and expand human potential in the future. The relevance of genetics to the intricate unfolding and operation of human behaviour is far from being clear, and attempts to link specific genes to specific behaviours or behavioural capacities are frequently simplistic.

A single example of neurogenetic intervention in laboratory animals and its consequences will illustrate how complicated all this is. A key brain property is that of plasticity – that is its capacity to modify its structure and cellular connections (synapses) in response to experience, in which one molecule known as the NMDA receptor, plays a major role. One subcomponent of this protein, the NMDA receptor subunit 2B (NR2B), has been particularly implicated in such changes in learning and memory. It was reported that increasing the quantity of the NR2B protein in the synapse by overexpressing its gene in a mouse results in improved learning and memory (Tang *et al.*, 1999).

The authors – and even more the publicity that followed the publication – concluded that this 'smart mouse' pointed the way to genetic improvement of intelligence in humans. It was, however, reported soon afterwards that this overexpression has many other effects as well, notably an enhancement of inflammatory pain, i.e. persistent or chronic pain in response to tissue injury and inflammation (Wei *et al.*, 2001). As a matter of fact, the involvement in the pain response has identified the NR2B molecule as a potential target for novel specific analgesics (Chizh *et al.*, 2001). In a way, the super-mouse (as it was dubbed in the enthusiastic journalistic coverage of the original scientific report) became a modern epitome of the biblical aphorism '. . . and he that increases knowledge increases sorrow' (Ecclesiastes 1, 18).

The NR2B story echoes, not in the details but in the principles involved, a much earlier attempt to isolate single gene mutations that affect learning and memory in the fruit fly, *Drosophila*. Though several very interesting and useful mutants have indeed been isolated, they do modify the activity of other behaviours and other body systems as well; the idea that we are bound to identify illusory 'data convertase' or 'dihydromemorase' proteins, that affect learning or memory alone, has long been abandoned (Dudai, 2002). Interesting as the specifics of the NR2B story are, its main value is as an illustration of the complications that can follow when one gene product is manipulated with the aim of altering behaviour without taking into account the complex function of this gene *in situ*. Such genetic manipulation would better be regarded as an exploration of function than as a targeted modification of function.

Misconstruction of the evidence, though not always by the original investigators, often generates superfluous expectations of neurogenetics. It has been suggested by some authors (Benjamin *et al.*, 1996; Ebstein *et al.*, 1996), though questioned by others (Gelerenter *et al.*, 1997), that novelty-seeking in humans is contributed by a certain subtype of receptor for the neurotransmitter dopamine. Even if this conclusion were to be proven valid, the evidence suggests that the influences are only minor, and that in any case the same subtype affects other behaviours as well. Tinkering with the composition of receptors to augment novelty-seeking might therefore culminate in anything from lack of appreciable effect on novelty-seeking to other, motor, emotional or cognitive alterations.

BRAIN TRANSPLANTATION

Repair of brain dysfunction by transplantation of brain tissue is an idea that has undergone multiple transformations. High hopes were no doubt reinforced by the remarkable success of organ transplantation in general, although nobody seriously contemplates transplantation of a whole brain, as if it were a kidney or heart (except in some horror movies or in jokes made about state leaders). There are some instances in which the idea could make sense. Consider Parkinson's disease which involves degeneration of brainstem nuclei that produce the neurotransmitter dopamine; why not introduce into the brain neuronal tissue that produces dopamine? In other cases, the aetiology of the disease and the relevance of the proposed solution are much less clear. The idea of transplanting neurons that produce the neurotransmitter acetylcholine into the brain of patients with Alzheimer's disease is based on the hypothesis of dementia that blames loss of acetylcholine for the dementia, when it is actually not yet clear whether this is the cause or the effect or even a primary contributor to the pathology; and the idea of implanting neural tissue into the brain of schizophrenics was merely a shot in the dark.

However, even in those cases in which the rationale is sound (Parkinson's disease), we have so little information on the fate of transplanted tissue that the justification of experimentation on humans becomes seriously questionable. The introduction of fetal brain tissue or porcine brain tissue into adult brains with Parkinson's disease was explored as much as two to three decades ago, both with (e.g. in Sweden) and without (e.g. in Mexico) official approval, but with disappointing results. Some leading experts now express enthusiasm for the use of stem cell technology, a trendy focus of attention in biotechnology in general, as a tool for brain repair; others promote the idea but sound cautious and realistic (see the Chapters 12 and 13, this volume). The questions are numerous: will the grafted cells acquire the identity and specialisation essential for fulfilling the role for which they are introduced? How will these cells integrate with existing circuits in the brain and affect brain functions that are unrelated to the targeted impaired function? Will the grafted cells go astray and invade other parts of the brain? Will they become malignant over time? And, as noted above in the context of the cholinergic hypothesis of dementia – even if the graft delivers the merchandise, is the targeted cellular deficit the source of the disease? Many of these questions are considered further in specific detail by Helen Hodges and colleagues (Chapter 12, this volume).

ELECTROMAGNETIC INTERVENTION

A technique that has recently attracted many investigators as well as clinicians is transcranial magnetic stimulation (TMS) – focussing a strong magnetic field across a specific brain region (Grafman and Wassermann, 1999). It is conceptualised as a 'virtual lesion' technique, which disrupts organised brain activity. TMS could be used in three types of applications. One is in basic brain research; after perturbing electrical activity in targeted brain areas, the experimenter observes any resulting behavioural effect, from which one could infer the function of the specific brain region. In this context TMS does not differ in principle from any other lesion technique, other than in that it makes a lesion that is considered to be functional, transient and reversible rather than structural. The second application is clinical. Perturbation of activity in a specific brain target could be explored for therapeutic ends, even if the mechanism of action is yet unknown – the prime example is the potential use of TMS in treatment of depression, bipolar disorder, obsessive–compulsive disorder, post-traumatic stress disorder and schizophrenia (e.g., Burt *et al.*, 2002). In the treatment of depression, for example, TMS is considered by some as an alternative to electroconvulsive shock therapy (ECT). The third potential application is direct stimulation of damaged brain areas to overcome sensorimotor deficits.

All in all, TMS is considered today to be of great potential. A major assumption is that the protocols used result in transient and reversible perturbation of brain activity. Studies have indeed been published to demonstrate the safety of this approach in humans. The question, however, remains whether the technique is as safe as promoted, particularly when applied repeatedly to the same target (repetitive TMS, rTMS). Aren't there any lingering lesions? Long-term studies, including in animals, are required to establish safety in the longer term and to define the balance between benefit and risk in specific cases.

This is an appropriate point to raise issues related to another physical technique, functional magnetic resonance imaging (fMRI), which is extensively used both in the laboratory and in the clinic. There is ample evidence to support the safety of magnets currently used routinely in the clinic and in most laboratories that conduct research on humans (having field strengths up to 3 tesla). Since higher-intensity magnets would provide even more powerful research tools, however, magnets of 7 T, and even 11 T, have now been introduced in a few research centres. There are multiple reports, mostly anecdotal, that the high-intensity magnets cause physiological and behavioural side effects. This must be further evaluated before human volunteers – often students – are subjected to such high fields, particularly if this is done repeatedly. It is noteworthy that rats avoid high-strength magnets – perhaps because they are smarter in sensing the risk than those who experiment on them!

ON COMMON FALLACIES

The application of risky experimental methodologies to the human brain is driven by several incentives, primarily of course a genuine urge to relieve suffering when there is no better treatment. Other drives may be less altruistic, however, and these together with several conceptual fallacies might explain why it is that even knowledgeable and critical scientists occasionally gravitate toward support for irreversible interventions in the human brain. These conceptual issues are particularly pertinent to the discussion of the potential use of the available or predicted outcome of recent neurobiological research.

The level fallacy

This is the assumption that once the system's components are identified, the system's function is bound to become understood. The revolution in modern biology is so far mostly due to the reductionist approach. This implies first and foremost 'constitutive reductionism', i.e. attempts to identify the molecular and cellular constituents of organisms. Constitutive reductionism is usually accompanied by the hope that 'explanatory reductionism' will ensue, i.e. that the knowledge of the components will explain the properties of the system as a whole (see chapters by Mary Midgley (Chapter 1) and Pat Bateson (Chapter 9); also Dudai, 2002). Constitutive reductionism is definitely fruitful, explanatory reductionism sometimes is. One area in which constitutive reductionism has been remarkably successful has been in analysing the chemistry of life, for example in the achievements of the Human Genome Project.

Much effort is currently funnelled into the study of interactions among molecules and molecular complexes at the cellular level, but conceptual understanding still lags far behind the accumulation of data. We find ourselves more and more at a loss in attempting to understand the interactions of molecules with each other in the cell. At the level of the research discipline, this calls for rethinking of the education of new biologists, with proper introduction of substantial elements of mathematics and engineering. At the level of research goals, it calls for rethinking of the timescales over which they can be achieved. Those who ever believed that deciphering the molecular details of the human genome would explain how the body works – if there were ever leading scientists who genuinely believed so – now encounter the need to delve into proteomics on the one hand and system theory and biological engineering on the other, in order to make sense of the data.

The problem is actually simple: whereas identification and isolation of the constituents of a system is a valuable step on the road to understanding the system, it is clearly not sufficient for such understanding. A useful way to think about this problem is to consider the classic analysis of levels promoted by Marr (1982), which applies to the operation of information processing machines in general. It discerns three major levels: (a) the level of the computational theory, involving the goals of the computations of the machine and the strategy to carry them out; (b) the level of algorithms, i.e. how the computations can be implemented in the machine in terms of input and output representations and of the transformation of input into output; and (c) the level of hardware implementation, i.e. how the representations and algorithms are implemented in the material from which the machine is constructed. The bona fide reductionistic approach may proceed top–down, from the computational to the algorithmic and then to the implementational level; but in practice, it seldom does. Investigators often proceed from a brief phenomenological description of the system to the implementational level, without really knowing what are the algorithms, and even more importantly, what is the computational goal of the system, with the risk of confusing the understanding of hardware (i.e. implementational level) with the understanding of algorithms and computations.

The 'ceteris paribus' fallacy

A powerful assumption in most research programmes is the 'ceteris paribus' assumption (Latin for 'other things being equal'). Such research programmes follow, almost by definition, a principle aptly formulated by Johnson (1751): 'Divide and conquer is . . . equally just in science and in policy.' Indeed in approaching a complex system, it makes much sense to focus on selected elements of the problem, preferably on a single element and, for the sake of analysis or argument, to regard the other elements in the system as unchanging and therefore irrelevant. Hence if one approaches a brain circuit, it makes sense to focus first on one type of synapse, and if one approaches synaptic function, to focus on one component of the synapse, etc. It is difficult to see, for example, how the systematic research programme on learning mechanisms in the sea hare, *Aplysia*, would have resulted in such a remarkable success, leading to a Nobel Prize for one of its instigators, Eric Kandel, had it not been for the initial intentional focus on only very limited processes in a sensory-to-motor type of synapse in the abdominal ganglion. However, at each stage of the analysis, we must keep in mind that ceteris paribus is usually a gigantic simplification. Neglecting this might easily lead to the rather embarrassing belief that the component explains the system.

A prominent example in brain research is a phenomenon called long-term potentiation (LTP), an artificially induced change in the electrical properties of synapses that many consider to be either a model or a mechanism of the brain plasticity involved in learning and memory formation which will be further discussed under 'The model/metaphor fallacy' below. For now suffice it to say that even if one assumes that LTP is indeed a neuronal mechanism of learning and memory in the brain, clearly it is not the only mechanism. Studying it in isolation is often productive, but as far as the neurobiology of memory is considered, one should never lose sight of the existence of other mechanisms that must interact with LTP in real life to generate the lasting memory. Or, if we go back to the NR2B case, even if we understand all the nuts and bolts of the NMDA receptor protein, and convince ourselves and our colleagues that NR2B is indeed essential for memory, tinkering with it will affect other proteins and cellular systems as well.

The model/metaphor fallacy

Models are either abstract or concrete systems that represent other, usually more complex or less familiar systems, or schematic representations that account for properties of systems. They are useful in attempting to explain and predict as yet unknown, complex systems. Models are either mathematical, or diagrammatic (in which case they are often metaphors, see below), or 'simple systems' of which I discuss two particular types. One is the use of apparently 'simple' organisms to investigate phenomena and mechanisms that are difficult to approach in more complex systems, for example, the use of *Aplysia* or *Drosophila* or the honeybee to investigate how animal behaviour is conditioned. The other is the use of simplified preparations, such as cell cultures or brain slices, or isolated phenomena, such as LTP, to investigate more complex phenomena such as real-life neuronal circuits (for instance tissues and cells maintained in culture) or learning and memory (LTP). Models must never be equated with the system they are assumed to model; they are at most partial analogues. Even if we understood LTP, we would not, contrary to assumptions that are apparently widespread, understand everything about learning and memory; and the behavioural aberrations in a transgenic mouse are not to be be equated with those of a demented human patient.

An example of an influential conceptual–diagrammatic model in brain research is the Hitch–Baddeley model of working memory – that is, memory systems activated by recall (Baddeley, 2000). This model depicts working memory as two 'slave' systems controlled by a limited capacity central executive. One system, the 'phonological loop', is specialised for processing language, whereas the other, the 'visuospatial sketch pad', is concerned with visuospatial memory along with an 'episodic buffer' to bind temporary episodic representations. Neurobiological studies of brain substrates of working memory often refer to the Hitch–Baddeley model not only as a pattern of generalised relationships but also as a system of components for which biology has only to seek the circuit, cellular and molecular counterparts. In spite of its effectiveness, it is still a model, and even when neurobiological data are found to be congruent with it, it does not follow that we have reached understanding of the real working memory system in the brain – assuming that such entity is at all honoured by the brain itself.

The 'metaphor' part of the model/metaphor fallacy refers to the more general conceptual framework of brain research rather than to specific models. We universally use metaphors to explain the yet unexplained, including the counter-intuitive (see also Chapter 4, this volume). It is even argued that we cannot dispose of metaphors in relating to the world, particularly to those segments of the world that are not perceived directly without instruments such as a particle accelerator, microscope or a telescope. Brain research has its own wide repertoire of metaphors, which are themselves drawn from contemporary culture and technology. Hence brain scholars in antiquity contemplated wax tablets, in the Middle Ages storage cabinets and palaces, in the Renaissance hydraulic systems, in the mid twentieth century operational control systems, and in the late twentieth century electronics and computers. Though very helpful, metaphors, like working models, should be treated as no more than stepping stones to better understanding. The investigator must not fall into the trap of believing that the metaphor has sufficient explanatory power to account for reality, so confusing the two. Further, even when the metaphor is understood as not have sufficient explanatory power, some attributes may leak from the metaphorical domain to the real world target domain, and lure the investigator into false parallels. Hence storage metaphors may lure us to think about memories as static entities, computer metaphors as segments on hard disk. The Hitch–Baddeley working memory model, mentioned above, involves metaphors such as a central executive and sketchpads. It is clearly unwise to attempt to physically manipulate brains on the basis of metaphors alone. Even if these metaphors are powerful and stimulate fruitful research, new cultural and technological concepts may transform even the most appealing contemporary metaphors of today into historical anecdotes.

HOW WILL WE KNOW WHEN WE GET THERE?

The title of this section is, admittedly, a rhetorical device. I have no criteria to offer, only a caveat. It seems to me that implicit in many disciplines of modern science and technology is a notion that has been crystallised by the eighteenth-century Italian scholar Giambattista Vico. He remarked that for the Latins, verum (the truth) and factum (what is made) were interchangeable. This was taken by him to mean that we can know nothing that we have not made. Though dating back to Christian scholasticism in the early Middle Ages, where it was an argument to support the ultimate knowledge of the supreme being, this maxim could be construed within the frame of thought of the secular mind and the world of natural and artefactual entities; it means that we have not achieved a genuine understanding of nature until we have developed the ability to reproduce or mimic it.[1]

Many brain scientists toy with the idea that constructing artificial brains or cognitive machines will explain how the brain works. It is indeed reasonable to conclude that if we build something we come closer to understanding it. Consider flight; since humans design and produce aeroplanes, they can consider themselves experts on aeroplanes. Once aeroplanes become airborne, the experience of the aeronautic engineer is unlikely to be sufficient to handle the consequences which will include issues of traffic control, effect of aeroplanes on culture, economy, environment, etc. – because aeroplanes do not operate in isolation, they interact with the world. Even if we end up producing primitive brain-like machines, this does not ensure that we will master the complexities of the human brain in action, let alone the complexities of the interaction of brains with their own bodies and with the brains and bodies of others.

The developments in the brain sciences over the past few decades have been remarkable and admirable. The excitements around the corner are bound to be even more impressive and nothing raised in this chapter belittles this view. My only aim is to emphasise the distinction between excellent science and the potential application of this science to modify human emotion, cognition and behaviour. My

[1] Interestingly, the 'verum-and-factum' maxim in its secular epistemological version echoes certain theories of brain action. These include motor theories of vocal recognition, which propose that to understand language we must have in our brain a neuronal machinery capable of producing language (Liberman *et al.*, 1967); and motor theories of action recognition, which claim that the observer matches its own motor representations of action with the percept to understand it (Jeannerod, 1994). But this is already another discussion which far exceeds the present one.

claim is that the level of knowledge and confidence in our understanding, which is required to permit persistent or irreversible modification of brain function (excluding of course life-threatening situations), far exceeds what we currently have, or in my opinion what we will have in the foreseeable future. This is not at all a mystical attitude, nor even a defeatist one, but simply anchored in the evidence. Humbleness and patience must be the name of the game.

11 On dissecting the genetic basis of behaviour and intelligence

ANGUS CLARKE

In this chapter, I argue that it is inappropriate at present to pursue research into the genetic basis of 'intelligence' and of other behavioural traits in humans. I do not think that such research should be prohibited, nor that we should ignore research findings that emerge from other studies and that give us insight into these areas, but I doubt the wisdom of conducting research designed specifically to identify 'genes' or 'genetic variation' that contributes substantially to the normal variation in human cognitive abilities and behaviours. Set out below are the various arguments that have brought me to this judgement, probably as much from temperament as deliberation. These considerations can be arranged on a variety of different levels:

1. First are a number of contextual issues such as what is intelligence? Why is it valued so highly? What is it that motivates some scientists to invest so much effort in attempts to measure intelligence, and especially to assess and rank their fellow humans? What lessons can be learned from previous attempts to measure the (intellectual and moral) worth of individuals and races/population groups?
2. Research aimed at identifying genetic variation associated with inter-individual differences in intelligence within the 'normal range' is relatively unlikely to yield important and replicable results and may consume much time, effort and resources.
3. Such research is unlikely to identify biological determinants of intelligence in 'normal' individuals or any clearly beneficial application in medicine or other social realms.

The New Brain Sciences: Perils and Prospects, ed. D. Rees and S. Rose.
Published by Cambridge University Press.

4. What is true of intelligence will be broadly true also of other personality characteristics, although the demarcation between 'normal' and 'abnormal' behaviours may be more difficult to define in some of these areas.
5. If genetic variation associated with variations in intelligence or behaviour is found, there could be serious adverse consequences regardless of whether or not such claims are eventually found to be valid. Some of these consequences may ensue even if no 'positive' results emerge at all – the very process of the research may have consequences.
6. And if genetic differences that are genuinely responsible for some of the inter-individual differences are found, there will be enormous social ramifications that would need to be thought through well in advance. For example, we can learn from current debates on genetic factors that may modify criminal responsibility. This chapter can be regarded as a contribution to this process of social debate.

I conclude that it is better to pursue behaviour genetics research in relation to clear impairments and progressive medical disorders. Such research will provide some information about 'normal' behaviours and intelligence but on aspects that will be more readily assimilated into and less disruptive for society. Here then is the argument in detail.

THE CONTESTABLE PRELIMINARIES

What is intelligence?

Is it even a single 'thing' or does the word refer to several distinct abilities? The community of human behavioural genetics enthusiasts has employed twin studies to demonstrate that much of the variation in IQ between individuals arises from differences in their genetic constitution, not just from differences in their environment and experiences. The worth of such studies can be challenged but there is no doubt that the heritability of IQ and many other psychological measures – the proportion of the variation in these measures within a given population that is accounted for by underlying genetic differences – is indeed substantial, although it will of course vary between populations and will depend upon historical and environmental circumstances. Heritability is not itself a biologically fixed constant. As Patrick Bateson has explained at length elsewhere in this volume (see Chapter 9), there are considerable conceptual difficulties inherent in disentangling the relative influences of genes and environment on any characteristics to which many genes contribute, and the notion that the genetic component can be isolated in the single measure known as heritability is deeply flawed. Further, the higher the selective pressure for a trait, the lower the heritability, because insofar as the trait has survival value, the more likely it is that those not possessing it will be selected against and the relevant genes eliminated.

It is interesting that studies of the heritability of intelligence have been used by behavioural geneticists to distinguish between two types of environmental factors, those shared and those not shared between individuals (Plomin *et al.*, 2001). Although this approach has not led to the identification of specific genetic or psychological mechanisms, it does provide evidence that siblings within a family can experience very different sets of environmental factors. For example, a child of high cognitive potential may elicit behaviour from his/her carers and companions that stimulates him/her more effectively, thereby allowing him/her to develop a still higher IQ. If the carers and companions are themselves of high ability then this process may be reinforced.

Another important issue to consider as we search for an understanding of intelligence is the general tendency of psychometric approaches to the study of mental life to treat their measurement scales as if they represented something concrete rather than recognising them for what they are – mental models in the heads of psychologists (see also Chapters 4 and 10). For example, it is easy to overlook the difference between 'intelligence' and 'IQ', and the appropriate distinction between the meaning of a word and the meaning of a measure that shares the same name; similar issues arise with many other psychometric variables, especially for measures of aspects of 'temperament' and 'character'.

Finally, we may ask what types of genetic variation in intelligence might be expected from present understanding of brain mechanisms.

Why is intelligence valued so highly?

Intelligence correlates strongly with wealth, social position and social 'success'. In the terms of sociobiology and of evolutionary psychology, it can be seen as a major determinant of fitness; it underlies the major components of behaviour that are relevant to Darwinian natural selection. In those terms, intelligence should be reducible to the behavioural traits that enhance the number of progeny left by an individual. There is no simple relationship between intelligence (as commonly understood) and fertility or fecundity, however, and indeed the behaviours that might be expected to maximise reproductive success will vary enormously with both the social organisation and material circumstances of the social group. Indeed, there are good grounds for expecting that a trait that is an important influence on fitness will generally demonstrate a low heritability, because selection will have been operating over so many generations. This is perhaps a counter-intuitive notion that requires some thought to be accepted when first encountered. Human behavioural characteristics that might be relevant to fertility include courage (but not usually to the point of self-sacrifice), political power and maturity of judgement, ardour, the willingness to laugh or to cry, the capacity for fun and wit, empathy and heartlessness, perseverance and fickleness, commitment and lack of commitment.

It is recognised that there are more males with intellectual difficulties and impairments than females, and an apparent excess of X chromosome loci at which mutations can cause such disorders.[1] This concentration of 'intelligence-related' genes on the X chromosome

[1] Males have only one X chromosome, females two.

would perhaps account for the greater variation in IQ among males than females, although it may not be justifiable to argue from the greater density of X chromosome loci associated with cognitive impairment to the greater density of loci at which normal (non-pathogenic) variation influences normal intelligence (Turner, 1996). One account of human evolution holds that the rapidity of the development of 'intelligence' in humans has been the result of sexual selection, with the tendency of females to choose the more gifted males leading to a runaway process of increasing intelligence – because intelligence became not only a factor directly involved in fitness, survival and success but also to be interpreted by mates as an indicator of fitness. There are other potential – alternative or additional – mechanisms at work, of course, such as the lengthening of the period of dependence in childhood and adolescence and the increase in lifespan in humans when compared with other primates, but they seem to attract less attention than the more glamorous model of sexual selection.

Clearly, the search for genes that maximise the behavioural component of fitness is going to be emotionally charged and also technically very difficult . . . but if it might lead to an improved understanding of crucial social terrain, then it could be tempting indeed. Paradoxically, there has been a long tradition of concern that the fecundity of those deemed by others as 'unfit to breed' might reduce the overall quality of the population, including the general level of intelligence. The readiness of some Darwinians to regard the lower social classes as less worthy of reproductive success than the more highly educated but less fertile does point to an independent source of social and moral judgements, sufficiently strong to override their commitment to the beneficial consequences of natural selection.

What is it then that motivates some scientists to invest so much effort in attempts to measure and explain intelligence? Some researchers might be attracted to this area because intelligence has connotations of social, political, financial and sexual success, justified by the authority of Charles Darwin and his cousin Francis

Galton. Who else would be more fit to wear the mantle of genius today than the successors of Darwin and Galton, pursuing the same ideas but with contemporary tools? An emotional commitment to the importance of heredity in determining intelligence is clearly evidenced by the fraudulent conduct of some investigators in the past and the abuses to which the measurement of IQ has been subject in relation to race, gender and social class. I remain worried by what underlies the vigour with which hereditarian views are sometimes promoted.

I would also argue that there has been a widespread overvaluation of 'intelligence', even to the extent of using it as the key item for assessing a person's quality. Other personal characteristics and behavioural traits may be put to one side to judge a person's overall worth by how 'bright' they are. This is certainly true in some institutional settings, but also more generally in society at large. This practice both reflects and reinforces an unhealthily narrow conception of 'the good life'. Limited intellectual capacity need not prevent a person from developing a sensitive and responsible character worthy of greater respect than many of the intellectually more gifted.

What lessons can be learned from previous attempts to measure the intellectual (and moral) worth of individuals and 'races'? The late Stephen Jay Gould tackled the conceptual flaws in the notion of *g*, the so-called 'general intelligence' constant, and many of the political and social abuses that have flowed from it, in his book *The Mismeasure of Man* (Gould, 1984). Given the perennial appeal of any approach that can be used to confirm social prejudice – such as a desire to discriminate against people of a different gender, 'race', class, language or social history – it is no surprise that the measurement of intelligence has assumed a prominent place in our social life and language. Thus can prejudices be confirmed – against unwelcome migrants, against any who come from a different part of the world or possess different physical features, against the poor, against women or men, or those from the neighbouring country, town or street. There is great potential here for mischief and tragedy.

RESEARCH INTO THE GENETIC BASIS OF HUMAN INTELLIGENCE IS UNLIKELY TO YIELD STATISTICALLY SIGNIFICANT RESULTS

Research into the genetic basis of differences in intelligence within the normal range is a search for genetic loci that contribute to population variation in a quantitative trait – the IQ. I have no doubt that genetic variation does influence IQ, but the idea that the variation between individuals at many different sites on the chromosomes will modify an individual's IQ in a simple, additive fashion, is deeply implausible. Single changes of major effect are of course important in the origin of intellectual disability (mental handicap), but are unlikely to account for much normal variation. Given that the genetic contribution to variation in IQ is likely to be heavily influenced by interactions between genes, the developing organism and the changing environment, then the prospect of identifying loci that contribute minor, interactive effects to the variation in IQ is small.

The search for loci contributing to complex disease traits in humans has been similarly disappointing. There have certainly been successes in finding rare genes having major effect in those who carry the disease-associated variation, as in some familial cancers and in heart disease associated with familial elevation of blood cholesterol. However, such subsets of the common, complex diseases often account for only about 5% of those affected, and there has been little success with the genetic dissection of the remaining 95% of cases in a range of important, complex diseases and psychiatric disorders.

One of the most contentious issues in molecular genetics research into complex diseases and traits is how to recognise a weak association between a genetic variant and a disease. The statistical criteria used to establish association are essentially arbitrary: the greater the number of markers to be assessed, the greater the level of 'noise' which there must necessarily be in the data and hence the more stringent the criteria have to be made to filter this out. Lower values of significance are frequently accepted if there are independent pointers to a particular locus. The experimental design is usually decided on pragmatic grounds such as how many possible candidate genes the research team could investigate adequately, rather than on the basis of objective principles.

A few interesting findings have emerged with families having genetic abnormalities, such as a chromosome rearrangement which seems to be associated with both schizophrenia and an abnormal pattern of electrical activity in the brain in response to flashes of light. Two specific genes have been implicated in major behavioural or cognitive problems in unusual families where a behaviour appears to be transmitted as a strongly genetic condition, the *MAOA* gene on the X chromosome associated with impulsivity and learning problems and the *FOXP2* gene at 7q31, associated with severe impairment of speech and language. However, no such mutations were found in these genes in other affected individuals. Other studies of language impairment and of dyslexia in specific populations have suggested relevant variation on particular chromosomes, although the precise genetic details have not been identified.

In contrast to these encouraging if preliminary results from children and families with clear disorders, very little has emerged from large studies of variation in IQ in the general population. One gene in which mutations cause microcephaly (a marked failure of brain growth) has been found not to be associated with variation in brain size or intelligence within the normal population. Another result suggested that variation at the IGF2R locus on chromosome 6 might influence IQ but this has not been supported in attempts to replicate the finding. A few chromosomal regions have been identified in the preliminary stages of a large, complex (and costly) statistical survey using nearly two thousand markers, but no firm findings have yet emerged.

EVEN WHEN A WEAK ASSOCIATION IS FOUND, THE BIOLOGICAL MECHANISMS WILL BE DIFFICULT TO IDENTIFY

The only reason for seeking to establish an association between a variation in behaviour or IQ and DNA sequence is to provide a starting point for understanding the mechanism. Simple models of causation would however be inadequate for a system in which as many genes are active as in the brain: as much as about half of the total genome is expressed there at some stage in development, and variation at one site often marks a substantial chunk of DNA within which there are other variants nearby. It is likely to be impossibly difficult to be certain which variation(s) if any are relevant to IQ or behaviour. Such approaches are therefore unlikely to identify genes that determine intelligence or behaviour in the normal range, or that would find beneficial application in medicine or other social realms.

These pessimistic conclusions are reinforced by a consideration of the type of genetic variation that would seem likely, from current understanding of brain mechanisms, to relate to variation in IQ or behaviour. The brain works by communication between nerve cells at the sites known as synapses which are known to readily form, atrophy and decay under circumstances that would be expected to influence the quality of cognition. Genes likely to be involved could include those coding for a wide range of neural growth factors (see Chapter 12) and the protein building-blocks of the synapse as well as the genes involved in the functioning of the chemical messengers (neurotransmitters) by which information is passed. It seems likely to me that the 'best' version at any genetic site is likely to depend upon the nature of those with which it interacts, since it is the system as a whole that is to be optimised, and moreover this is a system that operates at many levels from the molecule to the cell, circuit and beyond. Learning and cognition presumably rely upon a web of balances between many such processes. To complicate matters even further, it is well established that the range of factors associated with IQ extends to the quality of intrauterine life and birthweight, and so will not necessarily be confined to genes active principally in the brain. From an evolutionary point of view if the optimisation of cognitive processes gives substantial advantage in terms of natural selection, any mechanism in which the organism's cognitive abilities resulted from simple, additive interactions between independent influences over the long history of intense selection, would probably have eliminated most genetic variation for such traits within modern populations. This consideration also argues for variation at many loci in an arrangement that allows for a multiplicity of optimum solutions.

Even in less complex phenotypes than IQ, we can see how difficult such an analysis can be. Investigators have spent many years dissecting the biological processes that affect a range of blood lipid levels and contribute to susceptibility to heart disease. Many genes influence these processes, interacting with each other, with the environment and with the early life of the individual, even before birth. From studies of insects, it can be shown that genetic variation can be maintained in a population because one allele at a gene locus may be advantageous at one stage in the life cycle or in one sex while a different allele is superior in the other sex or at a different stage of life. Are we going to be any less complex than fruit flies? Even in the apparently simple field (known as 'inborn errors of metabolism') in which mutations in single genes affect the function of a single enzyme, it is becoming clear that there is no such thing as a 'simple' single gene disorder since mutations in single major genes lead to complex metabolic changes whose consequences cannot be predicted with any great confidence (Dipple and McCabe, 2000). These and similar findings support the reservations I expressed some years ago in relation to both the scientific validity and the potential misapplication of results from studies on genetic susceptibility to disease (Clarke, 1995, 1997a, b).

If we grant that differences in cognitive ability will be no less complex in their origin than these examples of simpler traits, then we can hold out little realistic prospect in the medium term of obtaining interpretable results from studies of the genetics of IQ.

THE PROBLEMS FOUND IN THE GENETIC DISSECTION OF INTELLIGENCE WILL ALSO BE FOUND FOR BEHAVIOUR

The considerations surrounding intelligence also apply to behavioural and personality traits, with two additional important observations.

First, a demarcation between 'normal' and 'abnormal' behaviours may be more difficult to sustain. Second, the study of 'behavioural phenotypes' associated with specific genetic disorders has grown rapidly over the past few years without leading to any clear progress in understanding yet.

If the distinction between normality and abnormality is more blurred for behavioural traits than for cognitive ability, then the distinction between the complex origins of variation in the normal range and the relatively simple influence of disruptive mutations in some examples of disability, may not apply to behaviour. While everyday behaviours may be described by neutral words such as novelty-seeking, harm avoidance and reward dependence, these can gather connotations of pathology and stigmatisation if associated with extreme examples in experimental studies such as in schizoid behaviour. Although some clearly genetic disorders do display typical patterns of behaviour, there is no reason to suppose that a given behaviour has the same cause in different individuals since superficially similar behaviours may result from very different mental processes (Karmiloff and Karmiloff-Smith, 2001). Nor may we assume a simple, direct causal link between the mutation causing the disorder and the behaviour itself (Finnegan, 1998). Clinical and descriptive research into the behaviour associated with genetic disorders is unlikely to lead to explanations of much normal behaviour.

ANNOUNCEMENTS OF GENES INFLUENCING INTELLIGENCE AND BEHAVIOUR CAN HAVE SERIOUS ADVERSE CONSEQUENCES (WHETHER OR NOT THE CLAIMS TURN OUT TO BE JUSTIFIED)

Any suggestion of a particular genetic basis for individual variation in intelligence will almost certainly lead to an interest in which populations may be supposed to be more or less favoured by their inheritance. Even without scientific justification and without any further measures of intelligence being made at all, prejudice could be fuelled to become socially and politically explosive (Harper, 1997), as indeed it has in some past examples. If it were found that one particular genetic variant, associated with a slightly higher or lower IQ in one population, was more or less common in another population, that finding could easily be abused even if a sensible interpretation was impossible. Similar potential for abuse could result from studies of behavioural traits such as aggression, criminal behaviour or a wide range of personality variables deemed more or less desirable.

Focussing the undoubted public interest in psychology onto the genetic basis of inter-individual differences is also likely to distort popular ideas about behavioural stereotypes and the appropriate social responses to difficult behaviour and criminality. Such inappropriate and unhelpful 'geneticisation' of human behaviour would be promoted whether or not any of the trumpeted findings turn out to have any validity.

GENETIC UNDERSTANDING OF INTELLIGENCE AND BEHAVIOUR WOULD RAISE SERIOUS QUESTIONS FOR SOCIAL POLICY

If genetic differences are ever found to be genuinely responsible, after all, for differences in behaviour or intelligence, the policy implications for education and the management of social and behavioural problems would be enormous. This is apparent if we consider current debates on psychiatric disorders that may affect criminal responsibility, for example in the chapters by Bateson (Chapter 9), McCall Smith (Chapter 6), Radford (Chapter 8), Sedley (Chapter 7) and Cornwell (Chapter 14). Should those predisposed to criminality through their mental illness be excused at least some of their responsibility, or should they be subject to preventive detention and incarcerated because they are thought likely to commit a serious crime even if they have never done so?

SO, WHAT DIRECTIONS MIGHT RESEARCH INTO THE GENETIC BASIS OF BEHAVIOUR USEFULLY TAKE?

Genetic research into the causes of behavioural and cognitive disorders in the abnormal as distinct from the normal range is important and potentially valuable because it addresses real problems for the lives of many sufferers. Even if it leads to no cures, it will provide explanations and help improve the management of disease. No doubt it will also illuminate some aspects of normal human behaviour and abilities, but with less likelihood of opening the way to abuse than research on normal characteristics that could be targets for social engineering.

Research into the behaviour of a range of non-human organisms is also likely to lead to important progress, as illustrated by successes in understanding the genetic basis of relatively simple behaviour in *Drosophila*, such as the 24-hour rhythm in activity and the response to gravity. The effects of mutation and gene interaction are much more readily interpreted in this simple organism in which genetics, though certainly complex, can be more readily defined and manipulated. The mouse can be used to focus on behaviours with more direct relevance for human behaviour, again with sufficient control of the experimental process for the prospect of rigorous interpretation (Plomin, 2001). This understanding of ourselves as biological entities will be outside the context of socially charged problems and abuses that it is so important to avoid.

CONCLUSION

The central problem that I have attempted to address in this chapter is the temptation to scientists and media alike to promote the demonstration of mere statistical associations between genotypes and particular patterns of behaviour as 'explanations' which inevitably carry connotations of social judgement. The Nuffield Council on Bioethics (2002) has considered similar issues at greater length and in correspondingly greater depth, and also recognised the possible negative social consequences of some genetic research into human behaviour. The conclusions are rather similar to those in this chapter and also highlight the limitations and dangers of genetic tests of cognitive potential or behavioural predisposition. It raises the possibility that such tests will be devised and marketed within the foreseeable future, and from a very cautious position on their likely validity, insists on a need for oversight and regulation. It recommends research into the social consequences of genetic testing for behaviours and cognitive ability and argues against any application to education or prenatal or pre-implantation genetic testing. It also recommends that any bodies funding genetic research on human behaviour take great care about the quality of the research and the public's perception of it.

If and when genuine understanding does emerge from the genetics of intelligence and behaviour, however it will raise a number of moral questions for which we will need to be prepared, hypothetical though they might be for some time to come. Some of the questions for us to dwell upon are:

How should we manage a predisposition to crime, violence and anti-social behaviour before it has manifested? And after a crime has been committed?

What are the consequences of testing young children to identify their likely developmental patterns of cognitive ability and their personality traits? When may such prophecies be unhelpfully self-fulfilling?

Is it helpful to direct children and adolescents – perhaps adults too – into career paths that will make use of their genetic predispositions?

How should society cope with individuals who choose not to take full advantage of socially acceptable, even expected, interventions to modify anti-social behaviours? And with intellectual disability and behaviour problems?

For what behavioural tendencies or ranges of cognitive ability is it reasonable to offer prenatal screening and the selective termination of pregnancies, or pre-implantation genetic diagnosis? Or to offer genetic enhancements?

12 Prospects and perils of stem cell repair of the central nervous system: a brief guide to current science

HELEN HODGES, IRIS REUTER AND HELEN PILCHER

WHAT ARE STEM CELLS?

Stem cells are a very special category of building-block in the human body, versatile in that they can not only divide to make copies of themselves but also turn into many mature final forms that no longer divide. For example, stem cells from blood or bone marrow can turn into nerve cells, and those from the brain can turn into blood. There is intense interest in medical applications to restore and renew body parts by inducing stem cell grafts to multiply into new types of tissue needed for repair. This is a particularly exciting prospect for diseases of brain degeneration which are presently incurable. This chapter explains the important concepts in simple terms and offers an account of the extent to which this promise is being realised in practice and of the hurdles that still remain. The chapter to follow considers the ethical issues raised by these actual and potential advances.

WHERE DO STEM CELLS COME FROM?

Stem cells are found in embryonic, fetal and adult brain and body. The fertilised egg is definitively 'totipotent', meaning that all other types of cell derive ultimately from it. As the embryo develops into a fetus, stem cells become progressively programmed to become specific

---

The New Brain Sciences: Perils and Prospects, ed. D. Rees and S. Rose.

cell types, and before their final evolution into mature non-dividing cells they are often called 'progenitor' or 'precursor' cells. Different varieties of stem cells are found throughout a human's development, and those from very early stage embryos still have the potential to form the hundreds of types of specialised cell and are therefore said to be 'multipotent'. However this traffic is not one way, and progenitor cells can revert to earlier forms, so they retain considerable flexibility.

Current UK legislation permits embryonic stem cells to be derived from early embryos up to 14 days after fertilisation. They are commonly taken from 'spare' embryos 'left over' from fertility treatment. A more controversial route is the creation of human embryos by the technique used to produce Dolly the sheep. In this procedure, the nuclear DNA from an adult cell is inserted into an unfertilised, denucleated egg. The resulting cell can then sometimes be coaxed to start dividing and turn into an embryo from which stem cells can be isolated. In principle, this so-called 'therapeutic cloning' offers the potential for patients to supply adult nuclei from their own cells in order to create stem cells perfectly matched for transplantation and tissue repair. It is commonly confused with 'reproductive cloning' which generates an embryo for implantation into the womb of a surrogate mother, producing a near-identical genetic copy of the original donor. Human reproductive cloning is illegal in the UK and many other countries.

Although the fetus is further along the developmental road than the embryo, it is still in a natural state of rapid growth and provides a particularly rich source of stem cells. Those from the brain are able to turn ('differentiate') into any of the three mature types – neurons (for transmitting information), astrocytes (for, amongst other things, supporting neurons) and oligodendrocytes (which wrap neurons in protective myelin membrane enabling electrical signals to travel quickly between cells). Such human stem cells can be isolated and cultured from aborted fetuses, typically from six to twelve weeks of gestational age, provided that the mother freely consents, and the procedure follows strict ethical guidelines.

The adult brain also has its complement of stem cells but these are largely quiescent. They increase in number after injury, suggesting some capacity for self-repair which might provide another therapeutic route if a means can be found to boost their activity. However, they are very few in number compared, for example, with the abundance of stem cells found in bone marrow, muscle or skin, all of which have greater capacity for healing. Many people see the use of adult stem cells as ethically more acceptable than the use of embryonic stem cells. They have the additional advantage of providing a perfect donor match if they can be multiplied in culture and grafted back into the body of the same individual. However, repair with adult stem cells is not guaranteed and their success seems at present to be less likely than with embryonic or fetal material.

PRIMARY FETAL GRAFTS IN CURRENT CLINICAL PROGRAMMES

To date almost all attempts to repair brain damage have used primary fetal grafts. Aborted fetuses, as distinct from early embryos, contain cells well down the differentiation pathways and these have been used to treat serious conditions where other forms of treatment have failed or have had adverse side effects. Encouraging results have been obtained for Parkinson's disease, which is characterised by loss of a subgroup of neurons located in a part of the midbrain known as the substantia nigra. These neurons secrete the neurotransmitter dopamine into the region known as the striatum, which is crucial for the control of body movement and balance. Dopamine-producing tissue dissected from the fetus, and grafted into the patient's striatum, supplies the missing chemical to reduce the symptoms of tremor and rigidity. However, the true effectiveness of these grafts is difficult to assess. In one recent study (Freed *et al.*, 2001), improvements were seen only in younger patients and others were actually made worse. Comparison of outcomes from different clinical trials is also hindered by different transplantation centres using different procedures for grafting and assessment. Double-blind placebo controls are often bypassed by surgeons who do not consider them ethical for reasons discussed in the penultimate section below. It is not surprising that over the ten years to 2000, only about 300 Parkinson patients have received grafts, in comparison with the 4 million sufferers worldwide (Lindvall, 2000; Lindvall and Hagell, 2000).

Apart from difficulty in assessing outcome, fetal grafts suffer from a variety of problems which hamper their widespread use as treatments. First, many consider that use of fetal tissue can never be ethically justifiable. Second, to obtain sufficient tissue, several donors are used (up to eight fetuses have been needed for some Parkinson patients) so that the ages of grafted cells will be variable. Multiple donors are not only difficult to coordinate, but the need for them means that no two patients receive identical grafts. The treatment cannot therefore be standardised, making the results difficult to interpret. In addition, only a small and variable proportion (4–6%) of cells in the graft survive transplantation. When fetal tissue is used for transplant surgery, it is also essential to ensure informed consent of the donor mother supported by appropriate counselling, and that fetal and maternal tissue are rigorously and rapidly screened for freedom from disease. These necessary safeguards add to the complexity of the procedure. For all of these reasons, but primarily because the procedure will never be ethically acceptable to opponents of abortion, it seems unlikely that fresh fetal grafts will develop into a mass treatment for neurodegenerative disease.

CELL LINES

Living cells prepared by dissociating human embryonic tissue can be maintained in cultures supported by a system of signalling molecules (known as growth factors) to provide the right environment and stimulus for them. A proportion of them can grow and divide into daughter cells which then divide again and again to generate 'lines' of cell types which turn out to have the properties of stem cells. Typically these form clusters that can survive and differentiate when grafted into animal brains, but these so-called 'epigenetic lines' have not been shown to produce any behavioural recovery in impaired animals.

Cell lines with more therapeutic promise can be produced by genetic engineering. An 'immortalising oncogene' can be introduced into cells to make them capable of indefinite division, so that individuals can be separated out to grow into colonies of identical cells, each to be evaluated for potential application. Transplantation of stem cell lines prepared in this way might offer a way to sidestep some of the problems associated with fetal grafts. For example, while multiple embryos are needed to provide primary tissue for a single transplant recipient, embryonic or fetal stem cell lines can be pushed through an indefinite number of cell divisions prior to grafting. Fetal dissections are then needed only to establish the initial cultures. In theory, enough cells from just one embryo could be generated to treat hundreds of thousands of patients. Every cell would be a carbon copy of its predecessor, and treatment could then be standardised from one patient to the next (see Gray *et al.* (1999) for further comparison of the advantages and disadvantages of fetal and stem cell grafts).

Other useful genes can, in theory, be put into cells prepared for transplantation, including those coding for growth factors, to enable cells to produce the relevant growth factor in the brain after transplantation. For example, cells modified to produce nerve growth factor (NGF) have been shown to enhance survival of vital neurons and promote behavioural recovery in animal models of Alzheimer's disease and ischaemia. Genetic disorders or neurotransmitter deficiencies could be targeted in the same way by cells carrying healthy genetic information. However, the further development of these approaches will need to pay attention to possible safety problems associated with their use and evaluate the extent to which such animal models really do mimic the disease in humans.

STEM CELL SAFETY

Overgrowth

Since stem cells have the inherent *potential* to increase their number through cell division, there is a danger that they might continue to multiply in their new environment after transplantation, and so give rise to tumours. Indeed, Isaacson's group (Bjorklund *et al.*, 2002) has

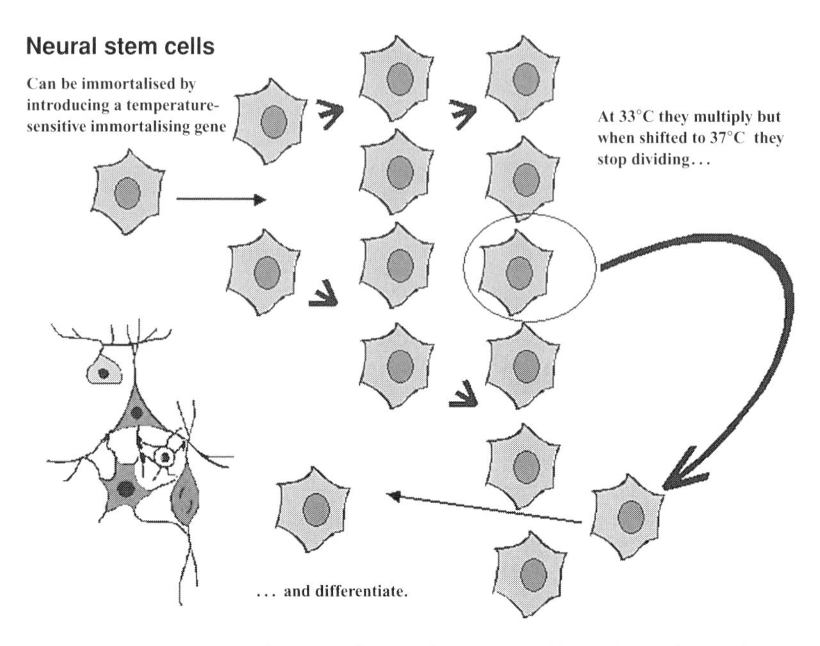

FIGURE 12.1. Schematic showing how stem cells can be made 'conditionally immortal'. A temperature-sensitive gene is inserted which allows cells to multiply indefinitely in a cool temperature, but switches off cell division at the higher temperature of the brain when they are grafted. Cells then follow instructions from the brain to develop into many different cell types.

found that tumours developed in five out of nineteen rats with surviving embryonic stem cell grafts in their brains, representing a level of risk that would be clinically unacceptable. This problem might be avoided by the use of 'genetic switches' to turn cell division on and off as required in the culture flask and the brain, which can indeed be done using the temperature sensitive properties shown by some genes. Cells to which such a gene has been transferred will divide at 33 °C but not at the higher temperature of 37 °C (either in the laboratory or when grafted into the brain). These cells are called 'conditionally immortal' since once grafted they are no longer capable of continued division at body temperature, and can then be induced to differentiate (Figure 12.1). It has even been possible to use genetic engineering to construct an entire animal in which all cells carry this temperature-sensitive gene, the so-called 'immortomouse' which provides a rich source of various conditionally immortal cell lines. Rats grafted with such cells have survived for periods of up to one year, with no sign of cell overgrowth or tumours (Veizovic *et al.*, 2001) but it has yet to be shown whether conditionally immortal cells of human origin will also be safe.

Fetal transplantation suffers from a similar risk because these grafts also contain stem cells with the capacity to divide, and indeed overgrowth is common in animal studies. Since the human brain is so large and primary fetal grafts generally have a poor survival rate, the problem in clinical practice is more usually one of too few rather than too many cells. Nevertheless, small regions of uncontrolled growth could disrupt rather than restore neural transmission, and this has been held responsible for the increase rather than decrease in movement disorders in some patients grafted for Parkinson's disease (Freed *et al.*, 2001).

Risks associated with viruses

Therapeutic transplants might carry viral infection in any or all of three possible ways:

- Transplants of human fetal tissue may harbour viruses such as human immunodeficiency virus (HIV), which can transmit from the donor to the recipient. Whilst the risk can be minimised by rigorous screening before implantation, this is obviously possible only for previously identified viruses. In addition, any methods for viral screening must be rapid so as to minimise the interval between tissue dissection and grafting. Viral screening cannot therefore be as rigorous or wide-ranging for fetal tissue as for cultured cells, although the checks can be extended by screening of the mother before taking the tissue.
- As stated previously, some cell lines are created by the introduction of a foreign, immortalising oncogene. These genes are commonly introduced into the cells by piggybacking on an inactivated virus of a type known as 'retrovirus' whose natural mechanism of infection is by

genetic modification (HIV is an example). In this instance, there is a theoretical risk (though generally reckoned to be negligible) that mutations may transform recipient cells to become tumorigenic. Primary cells harvested directly from embryos do not suffer from this potential problem but do have other disadvantages already mentioned.

- Non-human stem cells used as a transplant source (e.g. from pigs or mice) also carry a theoretical risk that endogenous non-human retroviruses could possibly 'jump species' and infect host tissue. In a worst-case scenario, this could cause permanent germline changes to subsequent generations. In a recent animal study, when insulin-producing pig pancreatic islet cells were transplanted into immune-deficient mice, pig endogenous retroviruses jumped the species barrier and infected mouse tissues (van der Laan *et al.*, 2000). However, it so happens that many conventional (non-stem cell) pig tissues have been transplanted into human subjects over the years, as treatments for, amongst other things, diabetes, liver disease, and disorders of the immune system. A rigorous screen of 160 recipients of such transplants (Paradis *et al.*, 1999) showed no evidence of retroviral infection. However, infection risks from non-human stem cells cannot yet be discounted, especially in the light of experiences with bovine spongiform encephalopathy (BSE).

Chromosomal instability

Unless we suffer from a rare genetic abnormality (such as Down's syndrome), the majority of cells in our bodies contains the same complement of 23 paired chromosomes. However, recent findings suggest that embryonic cells have variations in the number or make-up of chromosomes, and these change as they divide. The abnormal forms gradually decline, but a small and diverse population of variants can still be detected in adult neurons. Transplantation of embryonic stem cells therefore carries a risk of chromosomal instability which might be greater in cultured cells, as increased time in culture means an increased number of cell divisions, with a greater chance for the events to occur. Any cell lines for transplantation need to be monitored for the number and arrangement of chromosomes and their stability over time.

Immune rejection

Although the brain accepts 'foreign' cells more readily than other body parts, it will usually reject them eventually unless the immune system is suppressed. Patients with fetal grafts are typically treated for long periods with immunosuppressants and sometimes anti-inflammatory drugs. This is unpleasant, can lead to liver damage, and increases the likelihood of infection.

Unlike adult cells, stem cells do not carry the markers to signal their foreign status and they are therefore less likely to be rejected. We have found that mouse stem cells grafted into the brains of rats and marmosets are not rejected, even without immunosuppression (Gray *et al.*, 1999, 2000). It is possible that human stem cell grafts will not require extensive immunosuppressive treatment, which would be a considerable advantage.

WILL STEM CELLS BE EFFECTIVE?

Fetal grafts are generally implanted as small cell aggregates which tend to remain clumped at the site of implantation and make limited connections to the surrounding host brain (Gray *et al.*, 1999, 2000). Stem cell lines behave very differently and migrate widely within the brain to occupy the sites of damage and may even differentiate specifically into the cell types required for repair, following instructions from the surrounding brain.

For example, in a rat model for unilateral stroke, stem cell lines grafted into the undamaged side of the brain migrated to the damaged hemisphere whereas some cells grafted into the damaged side migrated to the intact side (Veizovic *et al.*, 2001). It seems that the transplant responded both to signals of direct damage, and to signals from brain regions attempting to compensate for damage. (Brain imaging studies of human stroke victims show the non-damaged side 'lighting up' with increased activity, as previously 'silent' pathways

(a)

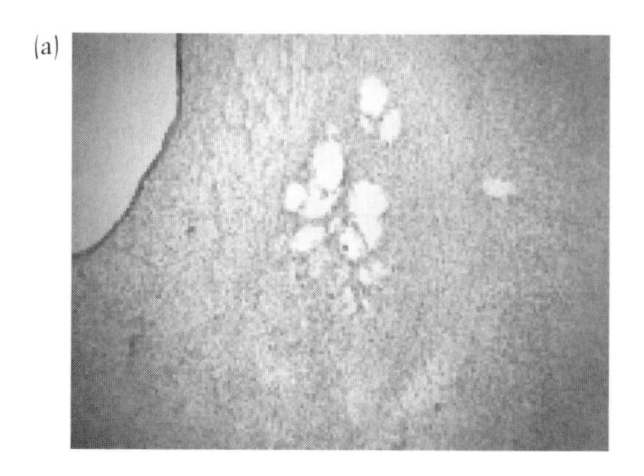

(b)

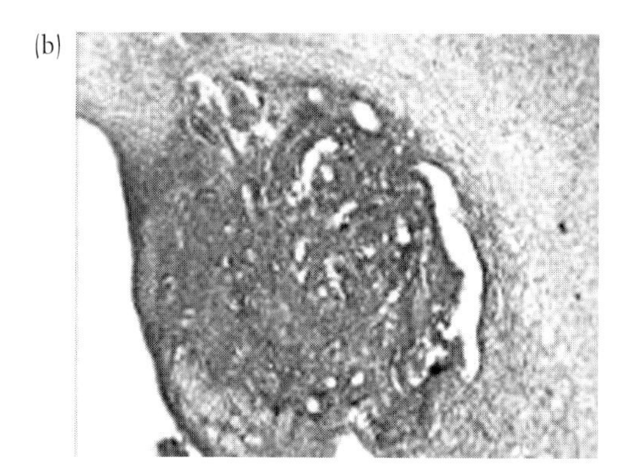

FIGURE 12.2. Stem cells grafted into the brains of rats with unilateral lesions of the striatum, a partial model for Huntington's disease. (a) Shows the lesioned striatum in a rat with 'sham' grafts, whereas lesion damage is substantially repaired in rats with stem cell grafts (b) of a 'conditionally immortal' cell line obtained from the immortomouse. (c) Lesioned rats with grafts spent as much time day by day as normal control rats searching in the correct quadrant for a safe platform submerged below the surface in a large pool of water. They also found the platform as quickly as controls. Lesion-only rats with 'sham' grafts, however, showed no improvement in search accuracy throughout 10 days of training, and rarely found the platform within the trial period of 60 seconds. (d) Rats with unilateral striatal lesions turn towards the lesioned side after injections of amphetamine, because of the increased dopamine release in the intact relative to the lesioned side of the brain. This 'rotation bias' was reduced at 8 weeks and normalised at 16 weeks after grafting with MHP36 stem cells. Bias remained high in lesion-only rats. ***Different from controls, $p < 0.001$; ••• different from lesion-only, $p < 0.001$; •• different from lesion-only, $p < 0.01$.

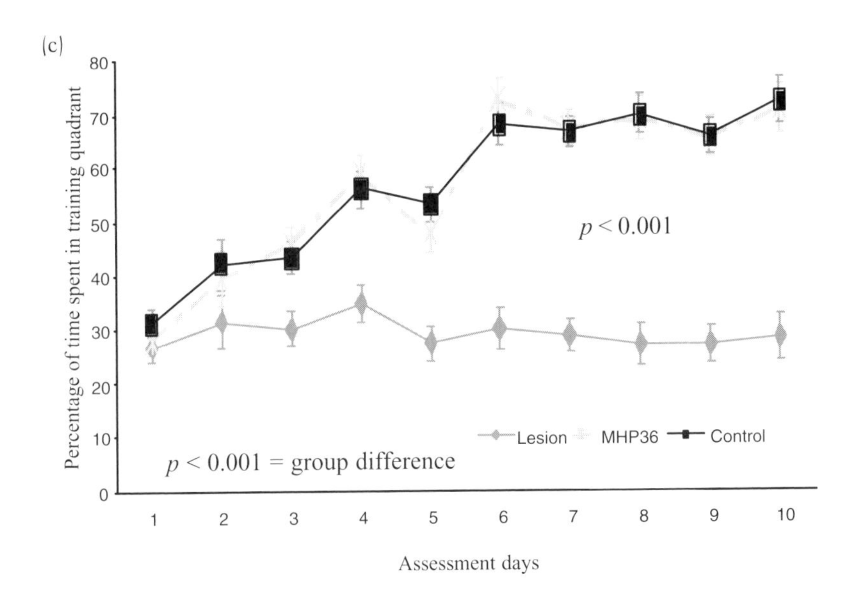

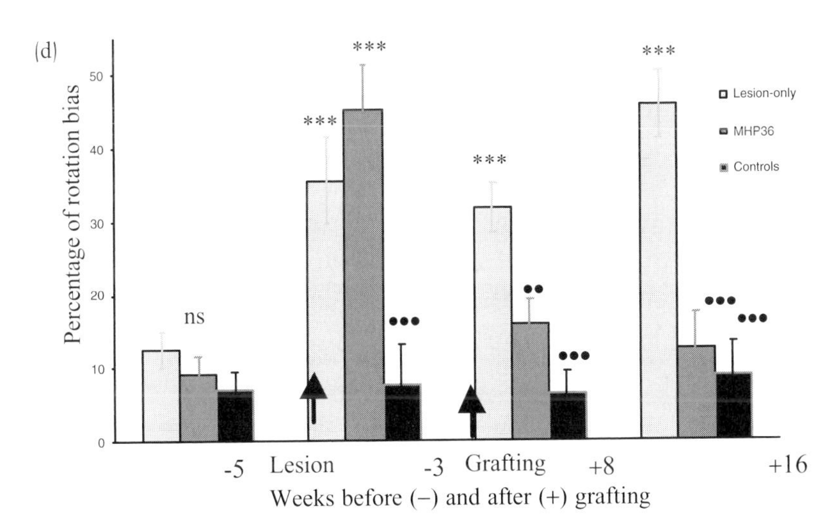

FIGURE 12.2. (*cont.*)

try to take on functions that have been lost on the side of the stroke.) In a rat model of heart attack in which blood flow to the brain is interrupted, there is early damage to the hippocampus, a region particularly sensitive to oxygen deprivation which is important for learning and memory, in which a specific cell layer is quickly destroyed. Stem cell lines grafted into the brain of these rats migrated to and specifically lined up along the damaged layer. With longer periods of oxygen deprivation, the damage was more severe and widespread and, correspondingly, the grafted cells dispersed more generally throughout the hippocampal region (Hodges *et al.*, 2000). Our recent results with a striatal lesion model of Huntington's disease show that stem cell grafts improved both cognitive (spatial learning) and motor (rotation bias) deficits, and accurately replaced missing neurons (see Figure 12.2).

These examples show that the capacity to migrate gives grafts of stem cell lines the ability to integrate much more naturally in the recipient brain than cells from fetal grafts and that when damage is limited they can merge into the structure and rebuild it. When the damage is more severe in some examples of stroke, grafted stem cells 'line' the area round the hole and merge into similar regions on the other side of the brain, possibly forming pathways parallel to those that have been destroyed (Veizovic *et al.*, 2001). This ability both to rebuild the architecture of the brain and/or to adapt it to new functions appears to be a very promising attribute of stem cells for repair. This very versatility might, however, have its downside from the regulatory point of view since it will be difficult to approve a therapy in which cell destinations and fates depend on the individual grafted brain and in which their wide dispersal makes tracking by brain imaging more difficult.

The moderately good record of success with mixed primary fetal grafts for improving or controlling the symptoms of patients with Parkinson's disease gives grounds for cautious optimism and the evidence so far on pure stem cell grafts suggests that they will prove to be as, if not more, effective. Some other cell types pre-differentiated prior to grafting have also shown promise in pre-clinical and clinical trials for stroke and Parkinson's disease.

Human neural stem cells have not yet been grafted into human patients, though differentiated lines have been tested to a limited extent (Borlongan *et al.*, 1998). It will be necessary to show safety and efficacy in animal models before they can be tested in clinical trials. Although tissue culture studies will be useful for clarifying the signalling mechanisms for differentiation, only transplantation into a living brain can reveal the extent of migration, the nature and extent of differentiation, the effects on the host brain, and the outcomes for behaviour and cognition.

When using fetal grafts, it seems important to implant the same type of cell as those lost, or at least as in Parkinson's disease, cells that release the appropriate chemical messenger. For example, rats with damage to the part of the hippocampus that guides the spatial learning needed for navigation through a maze, or finding a safe platform in a pool of water, show improved learning after grafts from the same area of the fetal brain but not after grafts from adjacent areas (discussed by Gray *et al.*, 1999). This limitation does not necessarily apply to transplants of stem cell lines, since we have found a single cell type to be effective in animal models of stroke, heart attack, Huntington's disease, Alzheimer's disease and ageing. Between them, these disorders represent different types of damage to diverse brain regions involving both cognitive and motor functions. However, this does not mean that all neural stem cell lines will be universally effective, since other lines derived similarly from the immortomouse were found to be much less effective. It seems that suitability for transplantation is under the control of particular gene sets, in interaction with particular types of damage. When better understood, factors underlying functionally successful repair should help us to predict which patient groups need which cell lines. Although animal studies therefore encourage the belief that stem cells will be an effective treatment for many different types of brain damage, there is still much to be learned about how they deliver their benefits in terms of the mechanisms underlying survival, migration and differentiation.

TOWARDS CLINICAL DEVELOPMENT

The need for effective treatments of neurodegenerative disease, stroke and brain injury is already overwhelming and accelerating with the increasing life expectancy of the population. For example, 5% of people aged 65 and 20% of those aged 85 will suffer from Alzheimer's disease. In principle, stem cell therapy could offer the potential for mass relief, with an improved quality of life for sufferers and their carers. Nevertheless, stem cell therapy is a moral maze. Even beyond the ethical questions about sources of stem cells and the risks to patients in experimental development, are dilemmas about priorities for their development and use. Will precedence be given to the severely demented to produce marginal improvements if these reduce the expense of care requirement even without significantly enhancing quality of life? Or will it be given to younger victims of stroke or traumatic brain injury even if they are coping relatively well, in the hope that they will recover their former abilities?

The possibility of 'cures for the incurable' has stimulated patient hope and demand on an unprecedented scale. Stem cell research laboratories are inundated with pleas for treatment and volunteers for clinical trials. In reality, the development of stem cells suitable for grafting is fraught with difficulties, with many major biological hurdles yet to be overcome. Even when stem cell therapies are assessed in human clinical trials, they may not fulfil the promise shown in pre-clinical animal studies. Such has been the case with many pharmacological treatments. Even so, it is worth bearing in mind that all of the patients who emerged from the Denver/New York Parkinson's disease transplant trial with worse rather than better symptoms of trembling (Freed *et al.*, 2001), said that the benefits were worth it – they would do it again. That is, the hope of improvement far outweighed the prospect of deterioration, so they would be prepared to take the risk of having another transplant.

Because stem cell therapy is novel, it poses severe problems for regulatory authorities faced with a baffling array of questions. Given the theoretical risk of viruses 'jumping species', would it be safe for recipients of non-human grafts to have children? Should the onset of clinical trials be accelerated for conditions, such as Huntington's disease, that are deeply distressing and without effective treatment? Is it really possible to obtain informed consent from those with chronic neurodegenerative conditions? What if pressure comes primarily from the carers rather than the patient – whose consent is authoritative now? Should grafting be risked when success is uncertain but the patient has little to lose, but on the other hand the public reaction to failure might prejudice the development of therapies for different conditions?

Assessment of patient recovery is problematic, especially when the comparison of 'before and after' conditions is complicated by continuing or uneven decline. Should treatment efficacy then be measured by actual improvement or by the slowing of an expected rate of decline? Although other types of treatment such as drug trials can use double-blind placebos to allow for the psychosomatic improvement which often results from powerful expectations, many surgeons are reluctant to use placebo controls for transplantation since it would involve giving patients sham surgery. Even if this problem were sidestepped by compensating the control group with promise of a later graft, the time interval for transplant assessment could well mean that their condition deteriorated beyond the point at which grafts might be successful. Effective assessment will also require the standardisation of procedures and comparison across centres. This is the goal of groups such as the Network of European CNS Transplantation and Restoration (NECTAR) (Boer, 2000), and the publication of Core Assessment Protocols for Intracerebral Transplantation (CAPIT) for Parkinson's and Huntington's diseases. These will involve matching and selecting patients very carefully on the basis of several types of information such as brain scans, history and thorough clinical assessments of symptoms over a period of months before surgery. Graft trials will be lengthy and complex.

Unlike drug levels which can be directly measured in blood or other body fluids to calibrate the dosage, the monitoring of grafted cells will have to rely heavily on proxy markers such as changes in glucose metabolism and blood flow in the brain, using imaging tools such as positron emission tomography (PET). The development of blood vessels in and around the graft can be followed by angiography, and the size and survival of grafts by magnetic resonance imaging (MRI). However the movement and merging of stem cells into the surrounding brain will make them ever harder to detect as the treatment proceeds, a problem that seems to have no solution with existing techniques. Chemical tagging can be used to highlight stem cells for MRI in animal studies but this would not be permissible in patients for safety reasons.

Huge research and development programmes will therefore be needed to bring the effective and safe use of stem cells to fruition, with correspondingly large expenditures. Such is the bright promise, however, that much corporate backing has already been forthcoming. Faced by the limited budgets and sometimes conservative attitudes of public funding bodies, stem cell researchers are increasingly seeking to patent their findings and develop their work through venture capital or funding from pharmaceutical companies. Academic and government administrations are also increasingly aware of intellectual property opportunities, all of which means that the development of stem cells is being nurtured in an environment full of commercial expectation and exploitation. Whilst this undoubtedly accelerates progress, the need to press on with product development might overtake the investigation of basic science issues essential to the understanding of graft mechanisms. The commercial interest and legal complexities of filing and protecting patents might also deter the progress and publication of research. More positively, privately funded biotech companies can often foster systematic, focussed and quality-controlled research, the like of which can be lacking in academic laboratories. Stem cell technology highlights a fundamental shift in the funding and development of science from the academic towards the private sector.

CONCLUSIONS

Stem cells provide a flexible approach to the repair of many different types of brain damage. However, clinical evidence for efficacy is lacking and functional improvement in animal models does not guarantee successful treatment for human conditions. There are major concerns about the derivation of tissue, and unknown risks associated with the therapy. Nevertheless stem cells offer the prospect of mass treatment for neurodegenerative conditions for which no alternative therapies are available or even in sight.

Further reading

The volumes edited by Dunnett and Bjorklund (2000) and Chadwick and Goode (2000) contain chapters by leading workers in the field of neural transplantation, describing methods and problems in both clinical and pre-clinical work, for example by Lindvall and co-workers on the Swedish experience with fetal transplantation for Parkinson's disease and by Ridley and Hodges and their respective co-workers on repair of hippocampal damage in rats and marmosets. A review by Bjorklund and Lindvall (2000) is balanced, brief and cautious, suggesting that transplants for Parkinson's disease are likely to be more successful than those for stroke, because the mechanisms of graft effects are better understood. Other reviews (Gray *et al.*, 1999, 2000) describe progress made with grafts of the MHP36 immortomouse-derived cell line in animal models of memory loss and other impairments. They contrast fetal and stem cell grafts and highlight diffuse and specific types of repair.

Amongst important original studies is the first demonstration that embryonic stem cells can become dopamine neurons and correct motor bias in lesioned rats, but with warning from five of the nineteen animals which developed tumours that a high proportion of stem cells, if unmodified, continue to divide after grafting (Bjorklund *et al.*, 2002). A tumour-derived cell line has been shown to differentiate into mature cells before grafting and to show some positive effects both in rats and people with stroke damage (Borlongan *et al.*, 1998). Stem cells migrating to both the damaged and undamaged sides of the brain have brought long-lasting improvements in motor and cognitive function (Veizovic *et al.*, 2001). A rare double-blind placebo controlled study found that only younger patients showed improvement after grafts of fetal dopamine-rich tissue, whilst movement disorder was worsened in five out of the thirty-three patients (Freed *et al.*, 2001). This study attracted much press criticism for exposing patients to risk, and some transplant programmes were temporarily halted. However, the patients themselves were very positive about the benefits of transplants, even those with poor outcome, and said that the treatment was worth the risk; they would volunteer again.

The role of NECTAR in coordinating transplant procedures across Europe is described by Boer (2000).

13 The use of human embryonic stem cells for research: an ethical evaluation

GUIDO DE WERT

The preceding chapter described how developments in the knowledge and potencies of stem cells are now holding out the promise of transplantation replacement therapies to restore organ functions that have been damaged or diseased, for example in Parkinson's disease, various types of heart disease and diabetes. Since these developments are currently drawing significant attention not only from biologists but also from the media, ethicists, governments, politicians, and indeed the general public, this chapter will explore the ethical issues causing concern. For further explanation of concepts and terminology, the reader is referred back to the previous chapter.

Much of the current ethical and societal debate is about spare embryos 'left over' from *in vitro* fertilisation (IVF) procedures, and the even more revolutionary alternative of embryos created specifically for the purpose by transfer of a cell nucleus from the patient's mature tissue (for example the skin) to a donor egg from which the nucleus has been removed. Cells from the latter would be 'autologous' (meaning: from the same organism) with the patient, holding out the promise of eliminating or at least substantially reducing the problem of graft rejection which normally bedevils transplantation surgery. This would have important benefits, for example for the sensitive human brain in the treatment of patients suffering from neurological or neurodegenerative disorders and handicaps, by improving on

The New Brain Sciences: Perils and Prospects, ed. D. Rees and S. Rose.

its natural advantage of suffering less violent immune reactions than other organs.

Some of the intense discussion has been more about cultural, philosophical and societal aspects of new developments in cell transplantation medicine, such as how we see ageing and chronic disease in terms of human values. Here, however, I restrict myself to more specific ethical questions about research using human embryonic stem cells, starting with the moral status of pre-implantation embryos and the ethics of using 'spare' IVF embryos as a source of human embryonic stem cells. I then briefly address the ethics of creating embryos for research into cell therapy by nuclear transfer, usually called therapeutic cloning (de Wert and Mummery, 2003).

THE MORAL STATUS OF THE PRE-IMPLANTATION EMBRYO

At one extreme of the spectrum of theories relevant to these questions are two positions which assert that the pre-implantation embryo is inviolable and sacrosanct: (a) the 'conceptionalist' view that the embryo is already a person, and (b) the 'strong' version of the potentiality argument – 'because of the potential of the embryo to develop into a person, the embryo has the same moral status as an actual person'. At the other extreme is the view that the embryo as a 'non-person' ought not to be attributed any moral status. In between these extreme views, there are various intermediate positions. Here, an 'overlapping consensus' consists: irrespective of many differences in the details of their arguments, these moderate views agree that the pre-implantation embryo deserves a real, but only relative protection, supported by (a) moderate versions of the potentiality argument, and (b) the so-called 'symbolic value' of the embryo: 'even though the embryo as such lacks moral value, we should not destroy it for trivial reasons because this would erode respect for human life generally'. These intermediate positions allow the 'instrumental' use of pre-implantation embryos provided that specific conditions are met, both procedural – informed consent of the providers of the sperm, egg and/or embryos and approval of the appropriate bodies charged with responsibility for oversight (usually national ethics committees) are absolute prerequisites – and material. Regarding the latter, there is presently a strong international consensus that embryos should not be used beyond two weeks of age (when the so-called primitive streak appears).

The international debate on embryo research concentrates on material conditions to be imposed with regard to the goals of the research and the source of the pre-implantation embryos. It is generally agreed that embryo research can only be justified by an important goal in relation to health interest. Opinions differ, however, on how this *principle of proportionality* should apply. A number of countries only allow embryo research relevant to human reproduction while others also accept cell replacement therapy as a legitimate aim. Different views also exist about sourcing from 'spare embryos' versus creating embryos specifically for research purposes. I will briefly analyse the various positions, starting from the principles to which most ethical arguments usually appeal.

ETHICS OF THE ISOLATION OF STEM CELLS FROM SPARE EMBRYOS

The use of spare pre-implantation embryos as sources of stem cells may be examined in terms of the principle of proportionality, the slippery slope argument and the principle of subsidiarity.

Principle of proportionality

It is difficult to defend the claim that isolating human embryonic stem cells for transplantation research is disproportionate if, as is accepted in many countries, spare embryos may be used for research into the causes or treatment of infertility. It is then inconsistent to reject research that may lead to the treatment of serious invalidating diseases. Given the major benefits that could flow this type of research, it is the selectivity of the ban that requires justification rather than its relaxation.

Slippery slope argument

The slippery slope objection runs as follows: 'isolating stem cells from spare embryos for research into cell replacement therapy is unacceptable because this may ultimately lead to genetic engineering of embryos for traits which could be fixed into further generations (germline gene therapy) and the manufacture of embryos for autologous transplants (therapeutic cloning)'. This objection is not convincing, for at least two reasons. First, it rests on the questionable premise that germline gene therapy and therapeutic cloning are categorically wrong: if germline gene therapy were to become a safe procedure in the future, it might well be morally justified, for example, in those rare cases where couples would only produce embryos carrying genes predisposing towards disease or disability and for whom therefore preimplantation diagnosis and selective transfer would not be an option. Second, even if germline gene therapy and/or therapeutic cloning were unacceptable, it does not follow that stem cell research for replacement therapy would lead automatically to its development.

Principle of subsidiarity

This principle means that the instrumental use of embryos can be justified only if suitable alternatives are not available for achieving the same ends. Critics claim at least three such alternatives, which have in common that they do not involve the instrumental use of embryos: (a) xenotransplantation – the use of transplants from animals; (b) stem cells from (dead) aborted fetuses; and (c) 'adult' stem cells. The question is not whether these possible alternatives require further research – this is uncontestable – but whether only these alternatives should be researched. Is a moratorium for isolating human embryonic stem cells for research into cell replacement therapy required, or is it preferable to investigate the different options in parallel? The answer to this question depends on how the principle of subsidiarity ought to be made operational. Although the principle of subsidiarity is meant to express concern for the (although limited) moral value of the embryo, it is a sign of ethical one-dimensionality to present every alternative which does not use embryos a priori as being morally superior. Any comparative ethical analysis of human embryonic stem cell research on the one hand and the possible alternatives on the other hand would need to take into account (a) the respective burdens and/or risks for the patient and his or her environment; (b) the corresponding benefits; and (c) the timescales expected to the clinical application of each.

In fact, there are practical and sometimes ethical questions to be asked about all the alternatives, not just about human embryonic stem cells. Xenotransplantation carries a risk of cross-species infections and a resulting threat to public health which represents, at least for the time being, an ethical barrier to clinical trials. It may also be asked whether it is ethical to breed and kill animals for transplants when there are spare human embryos that would otherwise be discarded. The use of cells from dead, aborted fetuses would seem – all other things being equal – more morally desirable than the instrumental use of living pre-implantation embryos, at least if pregnancy was not terminated for the specific purpose of obtaining the material for transplantation. For the time being, however, the fetal cells are very difficult to culture, with only one reported success. Studies with adult stem cells have given encouraging results, so much so that critics of the use of embryonic stem cells often argue that the adult cells have equal potential. Experts do not agree that this claim is yet adequately supported, however. Indeed, adult and embryonic stem cell researchers at an important recent workshop agreed that the versatility of adult stem cells had been overinterpreted and overplayed (Vastag, 2001) and a study for the UK Department of Health concluded that adult stem cells would only be brought to match the clinical promise of embryonic cells in the very long run, if at all (Department of Health, 2000). A moratorium which removed the best prospect for patients must be ethically questionable in itself. The way forward is surely through the evaluation of the different research strategies simultaneously. In any case, it seems that research on adult stem cells would succeed more quickly if it could take advantage of insights as they emerge from complementary work with human embryonic stem cells in parallel.

ETHICS OF THERAPEUTIC CLONING

Once again, I will explore the questions involved through an argument developed in stages.

A preliminary issue: creation of embryos for instrumental use

The moral legitimacy of this possibility was debated in relation to technologies for assisted reproduction even before the current controversies about therapeutic cloning arose. Two perspectives can be distinguished, namely a 'fetalist' focus on the moral value of the embryo, and a 'feminist' focus on the interests of women, particularly women as candidate egg donors. From a fetalist perspective it is often argued that, although the embryo is used instrumentally regardless of whether it is created for research or taken from a bank of spares, the degree of instrumentalisation is far greater for the former. Some commentators attempt to underline this difference further on the grounds of the intention at the moment of fertilisation. These arguments can be countered, however, by pointing out that any difference with regard to the degree of instrumentalisation is only relative if indeed it exists at all, that the moral status of each embryo must be identical (irrespective of whether it is a spare embryo or an embryo created for research), and that the true goal of regular IVF treatment is not in fact to create every embryo for its own sake but to solve the problem of involuntary childlessness in a procedure which anticipates and accepts the loss of some embryos at the outset. From a feminist perspective, it is argued that the creation of embryos for research involves the instrumental use of women to obtain egg cells by hormone treatment of them. Counter-arguments exist here too, however, in that the ethical principles are similar to those readily accepted for the recruitment of healthy volunteers to other medical research projects. Relevant questions are whether or not the research serves a sufficiently important goal, whether the burdens and risks to the subjects are (dis)proportional, and whether valid informed consent has been given. There is difference of opinion, also among female commentators, on whether giving hormone treatment in order to obtain research eggs may be considered (dis)proportional. No doubt, this problem may disappear when techniques of *in vitro* maturation are developed that will make it possible to avoid hormone treatment for egg donation.

The general conclusion is therefore the same from both feminist and fetalist perspectives, that there is no categorical objection to the bringing into existence of embryos for instrumental use.

The central issue: ethics of therapeutic cloning

To consider the moral acceptability of therapeutic cloning, I return again to questions of proportionality, slippery slope and subsidiarity.

Principle of proportionality

It is difficult to see how the principle of proportionality can provide a convincing a priori objection to therapeutic cloning when it is considered acceptable to create embryos for research aiming at the improvement of assisted reproduction, for example to improve cryopreservation and develop techniques for the maturation of eggs in culture.

Slippery slope argument

A consequentialist objection, fashioned as a slippery slope argument, is that therapeutic cloning will inevitably lead to reproductive cloning, that is the birth of human babies by means similar to those for Dolly the sheep. This objection firstly presumes that reproductive cloning is necessarily and categorically wrong, a premise still under debate. While there is general agreement that the serious health risks for children conceived by cloning make clinical trials on reproductive cloning premature if not criminally irresponsible at present, it is possible that somewhere in the future these risks could be controlled. Could reproductive cloning never then be justified for any case? Even if the answer were negative, we should realise that reproductive cloning need not follow automatically from research into therapeutic cloning. If recent claims that human reproductive cloning has already resulted in the birth of several children were to be substantiated, the current objection to therapeutic cloning might become even less convincing.

Principle of subsidiarity

The question for this stage of the development of medicine is not whether therapeutic cloning itself should be practised, but whether research should be allowed that might lead to it. Nobody believes that clinical trials of therapeutic cloning will be justified in the near future or indeed for some years to come, let alone that it should be introduced into routine clinical application. The need at the moment is for basic research, mainly to learn how to direct the differentiation of human embryonic stem cells. This research can, and should, be done with spare IVF embryos – it is not necessary to create embryos specially for this research purpose. At the same time, the principle of subsidiarity requires that research into possible 'embryo-saving' alternatives for therapeutic cloning be stimulated too. Amongst the lines of enquiry that have been proposed in the search for alternative procedures for autologous transplants are: (a) the use of adult stem cells; (b) transplanting a cell nucleus from mature (so-called somatic) human tissue into the enucleated egg of a non-human animal; and (c) direct reprogramming of adult cells to revert to former unspecialised states from which they can then be influenced to develop into other types of tissue (thus directing development without the need to create an embryo).

All these alternatives are more or less speculative. Any that develop real promise will need to be evaluated ethically as well as practically against each other and against human embryonic stem cells which remain the brightest prospect for the moment, though this too has yet to be demonstrated in practice.

In case therapeutic cloning becomes a viable option in the future, the question whether adult stem cells can really be an alternative should be reassessed on the basis of the data then available.

The logic behind transplanting the somatic nucleus of a patient into an enucleated animal egg is to evade the problem of the instrumental use of human embryos, because it is suggested that the so-created 'units' are not human embryos. Furthermore, an advantage of this approach would be that plenty of animal eggs are available – the feminist objection to creating embryos for research would, of course, not apply. It is not yet known whether this is a realistic option, whether human embryonic stem cells can be effectively obtained following this approach. Limited animal research so far has had disappointing results. Anyway, also here the risk of cross-species infections exists (as for xenotransplantation), although this risk may be extremely small, because the nuclear DNA of the animal is removed from the egg. Furthermore, a problem of classification arises: one may well doubt whether is it reasonable to argue that this 'special combination' cannot be considered equivalent to a human embryo. In any case, the recommendation to prohibit this strategy is premature (Department of Health, 2000).

Direct reprogramming of somatic cells is seen by some as the ultimate future alternative for therapeutic cloning. Its development has, however, a moral price: it requires somatic nucleus transplantation in order to create pre-implantation embryos for basic research. This research could, it is hoped, gain insight into the mechanisms by which an adult cell nucleus can be reprogrammed. Obviously, this type of human embryo research could only be acceptable after adequate animal research.

CONCLUSION

I have argued that research with human embryonic stem cells for the development of cell replacement therapy is not precluded by the moral status of pre-implantation embryos, and that the use of spare embryos for this purpose can be morally justified when tested against the principles of proportionality and subsidiarity, and the slippery slope argument. Other lines of research should of course be encouraged as well, especially if they might bypass the need for embryonic stem cells. The question as to whether therapeutic cloning should be allowed too only becomes acute (a) if research with spare IVF embryos suggests that usable transplants can be obtained *in vitro* out of human embryonic stem cells, and (b) if the theoretical alternatives for therapeutic cloning are less promising or need more time for development than is hoped for now. In that case, therapeutic cloning can be justified on the basis of both the principle of proportionality and the principle of subsidiarity.

14 The Prozac story

JOHN CORNWELL

My theme is how ideas in neuroscience – laboratory work, theory and seminar room discussions – land in our communities via pharmaceutical promotions, the media, print journalism and litigation; and how there is growing gulf between commonsense notions of responsibility, and a medicalised model of criminal behaviour.

In 1994 I was commissioned by a newspaper to investigate a case that connected a suicidal killer, the drug Prozac and a civil action for liability. Over a period of twelve weeks I attended a trial, in the United States, that involved arguments about impulse control, free will, the action of brain chemistry on human behaviour and, because the arguments were presented, public misunderstanding of current science (Cornwell, 1996).

On 14 September 1989, Joe Wesbecker, a forty-seven-year-old pressman returned to the printing factory Standard Gravure, his former place of work in Louisville, Kentucky, and shot twenty of his co-workers, killing eight, before committing suicide in front of the pressroom supervisor's office. It was discovered soon afterward that Wesbecker had been taking a course of the antidepressant Prozac. Thus Eli Lilly of Indianapolis, the manufacturer and distributor of the drug, became a prime target in a subsequent liability suit brought by the survivors and relatives of the dead. According to his psychiatrist, Dr Lee Coleman of Louisville, Wesbecker had been prescribed Prozac to alleviate depression related to workplace stress and his complaints of continuing unfair treatment by the management at Standard Gravure.

The New Brain Sciences: Perils and Prospects, ed. D. Rees and S. Rose.
Published by Cambridge University Press.

Plaintiffs' counsel argued that the drug had disrupted Wesbecker's impulse control to a point where he was not responsible for his actions. Lilly, they insisted, had knowingly manufactured and marketed a drug that undermined impulse control in certain individuals, and had failed to provide appropriate package warnings about contra-indications. Lilly on their side denied negligence in the testing and marketing of Prozac. Avoiding the neuropsychological descriptions of depression the company standardly uses to promote Prozac, the defence strategy focussed on Wesbecker as the product of a dysfunctional family that had suffered from hereditary mental illness in three generations. At the same time they represented their drug trial procedures as irreproachable and portrayed Prozac as a great success story, an example of the advances of neuroscience in recent years.

The purpose of jury verdicts in civil actions in the United States is to involve members of the community in decisions about how people and groups of people should behave towards each other. While the litigation was supposed to focus on liability for alleged negligence in drug trials and package warnings, the jury were invited to make judgements about Joe Wesbecker, as if he were a defendant in a criminal action. They were asked to make judgements about the state of his mind and behaviour, the operation of the normal and the abnormal brain, the effect of fluoxetine hydrochloride, or Prozac, on a man who was agitated and depressed with a 'psychoactive' component. It was repeatedly claimed by the defence lawyers that more time and expense had been expended on Joe Wesbecker's background than on any serial killer in US legal history (400 depositions). The jury's guides and mentors in these difficult areas were counsel for the defence and counsel for the plaintiffs, who had primed themselves to explain to the jury (only one of whom had been to college) how the brain works, supported by a cast of highly qualified and highly paid expert witnesses, including one of the inventors of Prozac, Dr Ray Fuller.

Early in the trial, Judge Potter remarked that he had come across a statement made by a neuroscientist that 'for every crooked thought there's a crooked molecule'. He was referring to a sentence in a passage of a deposition by Dr Ray Fuller, the inventor of Prozac, which had been read to the jury in Fuller's absence:

> The brain like other parts of the body, is made up of molecules. And the function of any part of the body involves molecular changes, so that when any part of the body is dysfunctional there would be some type of molecular change that would be occurring. Personality disorders would have disordered molecular processes going on. As Ralph Gerard has said: 'Behind every crooked thought, there lies a crooked molecule.'

The reference to Ralph Gerard, a distinguished but now long-deceased neuroscientist, reveals the symbiosis that operates between the brain sciences and the marketing arm of the pharmaceutical industry.

That symbiosis includes the media. In the spring of 1994, in an attempt to bring Prozac's celebrity status under control, Lilly's public relations department distributed an article from the *Saturday Evening Post* entitled 'Seeking the wizards of Prozac'. The author was one Tracy Thompson, a journalist whose thirty-year depression had been transformed by the drug. Her adulatory piece, distributed in marketing packs by Lilly to thousands of physicians throughout the United States, provides an insight into how Lilly wished Prozac, its inventors and its consumers to be viewed as the latest round of super-hype settled down. That it eventually seized the attention of the judge and stayed in his mind through the trial is not surprising. Thompson, the journalist, conjures up a metaphor characteristic of the new psychopharmacology:

> Serotonin is a neurotransmitter, a bicycle courier of the brain, shuttling electrical impulses from one nerve to another. With less serotonin being absorbed, more of it will be pedalling around up there, delivering tiny jolts of electricity from nerve to nerve. You could say this little green and white capsule is about to ratchet up

> the voltage in my head. And, in some way – no one knows precisely how – this will help me feel better. Happier. Without anxiety, able to take pleasure; pleasure in ordinary things. Sane.

Thompson reports that she had met Prozac's inventors Drs Moody, Wong and Fuller, who had, she notes, 'a genial pastoral demeanour – a bit like Sunday school teachers, or small-town family doctors'. Fuller had told Thompson that both Fuller and Wong 'shared a boyhood dream to become seminarians', but instead they had become priests of an altogether different kind: the dispensers of spiritual beneficence through their psychopharmacological creation. 'In his theology,' writes Thompson, 'the soul is a collection of cells, if a dauntingly complex one.' Fuller too, she says, quotes Gerard's comment, as one of his scientific predecessors, on crooked thoughts and crooked molecules. 'So,' she continues, 'are we just fooling ourselves to think we are able to reason our way out of despair? Is there a chemical for every sadness?'

Fuller's answer was that every sadness is, undeniably, chemical. He explained:

> There was an experiment in which damselfish were kept in a tank with only a transparent wall between some big predator fish. The damselfish had every reason to think they were about to be eaten. After a while, the serotonin levels in their brains showed a marked decrease. It's illustrative, in a crude way. Loss, anxiety, repeated rejection – things we experience do cause neurochemical changes in the brain.

Mr Paul Smith, counsel for the plaintiffs, who had evidently fed extensively on Lilly's promotional Prozac material, opened the trial by describing for the jury the operation of the brain and of serotonin.

With the aid of a chart, Smith gave the jurors their first lesson in neurobiology (the quotes come directly from the transcript):

> You have a presynaptic nerve end, and you have a postsynaptic nerve end. There is in fact an empty space between the two nerve ends of the two neurons. And you've got like a bridge or string or something and you've got a gap here. And what you have is, you have a neuroelectrical impulse with thought: thought being transmitted via electrical impulses that are picked up by these serotonin cells. The point is, to get the thought from this side to this side, and the way serotonin does it is by travelling across. The thought travels across here and it falls into a receptor site.

Moving on now to the purpose of Prozac, which he described as 'to stop all the serotonin being reabsorbed and lowering the level' he produces an image designed to put all this difficult science in its place:

> It's like you have a bucket and you've got a constant stream in that bucket and you've got a hole in the bottom of the bucket. There is a smaller hole at the end, so you've always got a level of liquid in that bucket, but if you stop up the hole and have the same amount of fluid coming in, you're going to increase the level of serotonin; the theory being if you increase the level of serotonin you increase the mood.

In response, the counsel for the defence, Joe Freeman, attempted to explain the brain and human behaviour in his own words:

> The brain is made up of many neurons or what we might call major lines through which messages are sent and received by what we call neurotransmitters. Now, there are a lot of them. There are many, many, many neurons and there are many, many, many neurotransmitters and they are of different shapes and kinds in terms of being received by the proper neuron.

He too told the jury that serotonin was 'used' in sending 'messages', and that a low level of serotonin 'causes' depression. Prozac, he said was invented by Lilly scientists to combat the problem caused by such lower levels. Eli Lilly, he said, had decided on developing a compound that would bring the neuron pump back into balance. Now, he too had an image. Joe Freeman said:

> I like to think about the neuron as the mother neuron or the neuron that creates the serotonin chemical that sends these messages to start with. They figured out a way to inhibit that pump from reuptake of the serotonin that was sending these messages and to balance out the system, but they left in place the sense like a mother has when the child is full, the sense that the neuron, a presynaptic neuron, had or that mother had when there was enough out there for her baby to eat or to get those messages – she stopped sending the food. They stopped sending more neurons. And over here on the other side they stopped demanding for more because there was enough there and there was a balance there and there was a help there for those people who had been so drastically sick.

What seems clear in these garbled accounts of brain chemistry and behaviour is the tendency towards what Steven Rose has termed 'reification': the drawing of an equivalence between thought and specific neurotransmitters, as well as the notion of the mind as a kind of machine or computer. Central to the argument in the early stages was a simplistic equivalence, moreover, between agency and 'impulse control'. As the trial unfolded the plaintiffs' counsel argued that by raising the level of serotonin to unpredictably high levels, the drug had destroyed what Smith called Wesbecker's 'impulse control mechanisms', thereby robbing him of his autonomy. Smith focussed on Wesbecker's behaviour as being zombie-like as he stalked the plant murdering his colleagues.

The defence, on the other hand, had started by maintaining that Wesbecker's killing spree was 'inevitable' – a result of inherited mental illness and a bad upbringing. But when that case began to founder under the weight of contradictory testimony, the defence lawyers constructed a new argument based on Wesbecker's *mens rea* – his status as a moral agent: he had goals, the defence lawyers and their expert witnesses reiterated, and he had intent. According to the defence, the murders were wholly caused by Wesbecker's unimpaired free will. Since he was a rational individual capable of making plans and choices, argued Lilly's trial lawyer, his ability to control his actions had not been affected by chemical intrusion; it was the interaction of nature and nurture that had led to Wesbecker's personality disorder, a disorder that left him nevertheless in control. He alone bore responsibility for his actions.

It is tempting to regard these final and apparently contradictory diagnoses – chemical determinism versus moral agency – as mere tactics in a civil dispute. After all, the goal of the plaintiffs was not so much to establish the exhaustive 'truth' of a complex set of events, as to assign blame and put a dollar value on the resulting damages. It is surely significant, however, that these versions of who, or what, was responsible for Wesbecker's actions were wholly lacking in a social dimension – the effect of workplace pressure, the antagonism and prejudice of fellow workers, the failure of a family doctor to detect tell-tale symptoms, or his failure to do more than prescribe Wesbecker pharmaceutical remedies.

As it happened, the jury of decent Louisville folk accepted Lilly's version of the story with very little agonising. After the trial, members of the jury conceded that the scientific dimension (involving weeks of neuroscientific exposition) had left them baffled, that they rejected the notion that heredity and environment had anything to do with Wesbecker's actions.

The importance of the story of this trial, it seems to me, is what little impact current neuroscience is exerting on traditional notions of *mens rea* in communities, that still, more or less, share views on individual responsibility. The lawyers' versions of even the most basic ideas in neuroscience were difficult for lay people to grasp, especially at the interface of science and human behaviour. It is not surprising therefore that the defence lawyers, and the jury, fell back with evident relief on a simple, durable, easily understood, view of commonsense agency. This resort to moral agency emerged at the end of the trial as a kind of disembodied *deus ex machina*, alienated and detached from the foregoing neuroscience – all the more detached since the scientific ideas had been crudely reductionistic and garbled. It was clear that the model of personhood presented at the trial's end, and accepted by the jury, was essentially Cartesian, a spooky stuff soul in a machine-like body–brain.

The story of neuroscience, as I have found myself reporting it, as a journalist, through 'The Decade of the Brain', is not exclusively about radical reductionism. There is a remarkable plurality of theories and hypotheses, at least at the level of popular exposition, and especially at the interface of human/social behaviour. It does seem, on the face of it that neuroscience is moving inexorably in the direction of a determinism that has long been familiar in philosophy of mind since David Hume. Philosophical scepticism about free will, however, has had no significant impact on the lawyers' approach to criminal responsibility: principally, one imagines, because Anglo-American culture has long accepted that determinism is nevertheless compatible with individual responsibility. The philosopher Galen Strawson has another way of putting it – that to act freely we need to believe in freedom.

But how much longer can commonsense notions of *mens rea* prevail under the influence of neuroscience and new genetics which draw an equivalence between medical conditions, pathology and the notion of most forms of criminal behaviour as treatable? In their popular book and TV series, *A Mind to Crime*, Anne Moir and David Jessel urge that if criminal acts are increasingly linked with biological disorder we can make things easier for ourselves by replacing traditional concepts of justice based on guilt and punishment with a medical model based on prevention, diagnosis and treatment (Moir and Jessel, 1995). Alison Abbott (2001) discusses brain imaging studies that allegedly reveal psychopathy. The neuropsychologist Antonio Damasio is quoted in the article as endorsing the notion, and as saying that 'it is clear that brain dysfunction can cause abnormal social behaviour, it is important for scientists to address the issue. The human mind is complex and refined, but scientists should not be afraid to think big.'

The persistence of traditional ideas about *mens rea*, alongside a growing conviction that criminal behaviour can be identified neuroscientifically and genetically, indicates that there is a widening gulf between the categories used by neuroscience and medicine, and the non-scientific categories used by the criminal justice system. The gulf has reached such proportions that it threatens to undermine the principle and purpose of civil actions – namely how members of a community expect people to behave towards each other. There is a widespread impression of the criminal justice system being lenient on crime, and growing calls for severe punishment of perpetrators. Victims, it is often said, suffer a life sentence while the actions of perpetrators are regarded as sociobiologically, genetically and environmentally determined. The impetus to empathise with the victims of crime does not easily square with the views of those who seek to medicalise the predicament of the criminal.

In the light of the paradox, David Hodgson, a Judge of the Supreme Court of New South Wales and a distinguished student of neuroscience, has called for a theory of retributive justice, which continues to give a rationally defensible account of moral agency, *mens rea*, while appealing to authentic scientific research that does not entirely abandon the notion of personal responsibility (Hodgson, 1991). Such a theory should not ignore the complex social and communitarian dimensions of human behaviour, which are the contexts in which we live and have our being, the contexts moreover in which a criminal and civil justice system has meaning. The challenge is for neuroscience and social science (including jurisprudence and ethics) to stay in conversation, rather than allowing themselves the luxury of a polarised sectarianism that is less likely to make us a race of zombies than to perpetuate a Cartesianism that violates the scientific advances of three centuries, while dishonouring the social reality of human nature.

15 Psychopharmacology at the interface between the market and the new biology

DAVID HEALY

INTRODUCTION: FROM ANXIETY TO DEPRESSION AND BACK

As portrayed in pharmaceutical company advertisements, the typical nervous problems seen in both psychiatry and general practice from the 1960s through to the early 1990s took the form of an anxious woman in her twenties or early middle years. The exhortation was to treat her with benzodiazepines (such as Valium), marketed as tranquillisers. In contrast, during this period, advertisements for antidepressants typically featured much older women. However, in the 1990s young or middle-aged women with nervous problems were portrayed in the advertisements as depressed, with the exhortation to treat these problems with selective serotonin reuptake inhibiting (SSRI) antidepressants such as Prozac. By the end of the 1990s anxiety seemed all but forgotten by the advertisers. But, post 11 September 2001, the 'typical' woman is once again likely to be viewed as anxious, with exhortations to treat her with SSRI drugs, which seem now to have become anxiolytics. What is happening here? Are the biological bases of nervous problems really changing so quickly, or is this a matter of marketing of available new drugs, along with changes in nomenclature and fashion? There is arguably more to this than just a matter of changing fashions in the labels we put on nervous problems. In the past decade or so, pharmaceutical companies have developed

The New Brain Sciences: Perils and Prospects, ed. D. Rees and S. Rose.
Published by Cambridge University Press.

abilities to change the very language we use to describe our most intimate experiences. For a fifty-year period extending through to the mid-1980s, the management of office-based or general practice nervous problems was coloured by psychodynamic thinking, stemming from the work of Freud, Jung and other pioneers in this field. This thinking led to a loose use of psychodynamic terms in everyday life – a psychobabble – that became part of the cultural air we breathe. This language had clear consequences for the way we perceived ourselves across a broad range of domains, from our notions of moral agency to the way we raised our children. Under the influence of the marketing of SSRIs, this psychobabble is rapidly being replaced with a biobabble that talks of lowered serotonin levels and the like, in a manner that is divorced from any scientific frame of reference. Superficial though such a language is, it can be expected to have consequences for a range of issues from notions of moral agency to the way we raise our children, and more generally for societal willingness to accept the products and claims of the biosciences. In this chapter I explore the role of marketing in these issues by looking at the marketing of psychotropic drugs for the mental distress found in the general community.

THE MARKETING OF THE ANXIETY DISORDERS

In 1980, the American Psychiatric Association published the third edition of its *Diagnostic and Statistical Manual* (DSM-III), which specifies and classifies the patterns of behaviour and forms of mental distress to be considered when making a diagnosis of depression, anxiety or other conditions (see also American Psychiatric Association, 1980; Chapter 16, this volume). This publication had a big impact on the management of nervous problems by the administration of drugs. With DSM-III, the vast pool of mental distress in the community, previously seen as anxiety neurosis, was broken up into a series of seemingly discrete entities – panic disorder, social phobia, post-traumatic stress disorder (PTSD), obsessive–compulsive disorder (OCD) and generalised anxiety disorder (GAD). In contrast the varieties of depression existing before DSM-III, such as vital as opposed to reactive depression or neurotic as opposed to psychotic depression that were previously regarded as distinct diagnostic categories, were collapsed into the single diagnosis of major depressive disorder.

Following the publication of DSM-III, the Upjohn pharmaceutical company began to market a new benzodiazepine, alprazolam, as a therapy for panic disorder. This appears to have been predicated on a calculation that panic disorder would be perceived as the most severe form of anxiety and demonstrations of efficacy for this condition would lead to alprazolam being used for all anxiety disorders, thereby pushing out competitor compounds (Sheahan, 2000).

Upjohn funded extensive trials of alprazolam and, following on from this research effort, a series of conferences and symposia on panic disorder. Many of the senior figures in world psychiatry were recruited to these clinical trials and participated in the subsequent meetings, which did a great deal to popularise the notion of panic disorder.

Within a few years, even in countries such as Britain, in which alprazolam did not become widely available, the concept of panic attacks had penetrated deeply into professional and popular consciousness. People who had previously understood their experiences in terms of anxiety or simply 'nerves', now described them in terms of panic attacks. States of tension or distress lasting anything from half an hour to hours were termed panic attacks by patients, even though DSM-III indicated that panic attacks usually last minutes at most. Comparable changes occurred in the fashion for diagnoses of social phobia, obsessive–compulsive disorder and post-traumatic stress disorder in the wake of pharmaceutical companies obtaining licences to market compounds for these disorders. To explore how this might happen, we turn to the marketing of SSRIs for depression in the 1990s.

THE RISE OF DEPRESSION

The targeting of these disease entities in what had formerly been the anxiety continuum was derailed in the mid-1980s by a developing problem with Valium and other benzodiazepines. In the early 1980s, voices began to raise concerns that these drugs caused physical dependence. This concern quickly became widespread and threats of legal actions loomed on the horizon for general practitioners and others prescribing tranquillisers.

At the same time a new group of drugs acting on the serotonin system were moving towards the market place. The first of these, buspirone, was targeted initially as a non-dependence producing treatment for anxiety. It failed. Neither general practitioners or patients were prepared to believe that there could be such a thing as a non-dependence producing 'tranquilliser'. The management of anxiety disorders with drug treatments had become a problem and these difficulties led directly to the development and marketing of the SSRIs as antidepressants.

This was a calculated strategy with considerable risks, given that it had not been possible to demonstrate that these new drugs worked for classical depression – the form of depression that was thought then to affect somewhat older people, which often led to hospital admission and possible treatment with electroconvulsive therapy. However, antidepressants were not thought to be dependence-producing, and for this reason marketing these new drugs as antidepressants had advantages. Furthermore, the benzodiazepines were widely perceived as agents that helped people adjust to stresses rather than as curing diseases and were therefore open to the possible criticism that it was inappropriate to use a chemical crutch to adjust to life's stresses. By contrast, severe depression was widely seen as involving biological abnormalities and the SSRIs came with a claim that these drugs corrected abnormal serotonin levels. It was possible to present SSRI antidepressants as correcting imbalances in brain chemistry in a manner that gave the appearances of treating a disease. Furthermore, if the fault was a biological lesion within the person, there was something of a moral onus on people to have such a lesion corrected, so that they could play their part in society. The taking of Prozac and other SSRIs in response to the serotonin deficiency model that emerged seemed almost as uncomplicated as taking vitamins for a vitamin deficiency.

It also seemed clear in the late 1980s that the companies making SSRIs could envisage a strategy for extending their use from depression to anxiety disorders (Healy, 1991). This possibility arose because while classic melancholia differs as clearly from anxiety attacks as Parkinson's disease differs from multiple sclerosis, many patients in the community display symptoms of both anxiety and depression. Physicians could be 'educated' to notice the depressive symptoms. This education led to change in the perceptions of both patients and physicians just as clearly as Upjohn's educational campaigns with alprazolam had done a decade before for anxiety and panic disorder. By the mid to late 1990s, patients presenting to physicians would typically complain that their moods fluctuated frequently, often at anything from half-hourly intervals to every few hours. Such rapid fluctuations of mood are incompatible with classic notions of depression, which involve persistent symptoms lasting for weeks or months rather than hours.

As a measure of what happened in the West during the 1990s, it is instructive to look at Japan and the rest of the non-Western world. In Japan the benzodiazepine dependence crisis did not happen, and therefore there were no SSRIs on the Japanese market through to the year 2000. Since then fluvoxamine has emerged for the treatment of obsessive–compulsive disorder and paroxetine for the treatment of social phobia. Against this background, depression remains in Japan what it was in the West up to the mid-1980s, a more severe but much rarer condition than anxiety, and a condition that primarily affects older people rather than all ages of the spectrum.

A key point in considering these shifts is to realise that changes in prescribing practices are not driven by good evidence. Half the clinical trials undertaken to bring the SSRIs on the market as 'antidepressants' failed to produce any improvement in the condition above placebo. What this means is that SSRI agents can be shown to do something, but this something falls a long way short of eradicating depression in the way that an antibiotic might eradicate an infection. There is, in other words, sufficient evidence to warrant these drugs being licensed as antidepressants, so that they can be used, but insufficient evidence to warrant a wholesale switch of treatment strategies in the expectation that clear health benefits for either individuals or communities will follow.

POST 11 SEPTEMBER PSYCHIATRY

If drug development had been proceeding as rapidly as both scientific and public expectations might have hoped, by 2000 pharmaceutical companies would have had a clutch of new compounds for nervousness. The plan was to market these as anxiolytics rather than antidepressants or tranquillisers. Companies believed that a decade after the benzodiazepines, a simple change of word from tranquilliser to anxiolytic was all that would be needed to remove the stigma of treating anxiety with drugs for both the public and the prescribers.

In the absence of any novel developments in drug therapies for nervous disorders, from the mid-1990s onwards companies redoubled their pursuit of anxiety indications for the SSRIs. Two of the new anxiety disorders carved out of anxiety neurosis by DSM-III were of interest – post-traumatic stress disorder, and generalised anxiety disorder. By the end of 1999, Pfizer had a licence for sertraline for the treatment of the former. In 2001, Glaxo SmithKline got licenses for paroxetine for both disorders and Wyeth for venlafaxine for generalised anxiety disorder. The stage was set for the relaunch of the SSRIs as anxiolytics.

The events of 11 September 2001 provided a perfect backdrop for any company wanting to market an anxiolytic. Following the bombing of the World Trade Center, there was much media debate on the anxious times in which we now lived, with articles appearing in a range of broadsheets and other periodicals outlining the criteria for generalised anxiety disorder and often making generous mentions of SSRIs. In the United States, there was in addition vigorous direct to consumer advertising of SSRIs for anxiety indications. The messages were that anxiety affected even more Americans than depression. Since then companies have spent approximately $100 million marketing SSRIs, primarily as anxiolytics, and sales of these drugs continue to rise apparently inexorably.

The advertising of sertraline, venlafaxine and paroxetine as anxiolytics show overlapping commercially important messages. The first of these is that SSRIs correct the chemical imbalance that underpins anxiety. This chemical imbalance is the very same lowering of serotonin that large proportions of the population, including many physicians, have for some years believed underpins depression.

The second commercially important message comes in invitations to 'ask your doctor about non-habit-forming paroxetine today'. This may also be put as follows: both SSRIs and benzodiazepines are effective for the treatment of anxiety disorders. Benzodiazepine drugs, however, should only be used for a limited period of time since they tend to produce dependence. SSRIs are not benzodiazepines. In contrast to this between-the-lines message, the patient information literature states unambiguously that these drugs are non-addictive and non-dependence-producing.

However, before the current SSRI drugs came onto the market, studies had been undertaken with them, in which it was clear that even healthy volunteers could become physically dependent to these drugs (see Chapter 14, this volume). As of 1997 Lilly, the makers of Prozac, had advertising campaigns stressing that Prozac reduced the risk of withdrawal syndromes compared with other SSRIs, such as paroxetine.

This case study of the dramatic successes pharmaceutical companies have had marketing the SSRIs seems to demonstrate either the extraordinary power of the propaganda techniques in current deployment in psychiatry, an almost 1984-like ability to mould both the public and the psychiatric mind, or the extreme credulousness of the physicians who prescribe psychotropic drugs, or some combination of these two. How are such transformations achieved?

MARKETING

Physicians seem to assume that the marketing of drugs involves little more than straightforward advertisements or pens or mugs with a drug name on it. When asked to rate the influences on them, physicians list marketing as the least influential and evidence as the most influential, especially evidence from randomised controlled trials.

In fact, the marketing departments of pharmaceutical companies are involved in the development of a drug right from its early testing in human volunteers. Their considerations influence the trials undertaken, the recruitment of triallists, the choice of consultants to and speakers for the company. The trials done are best regarded as marketing rather than scientific exercises, set up to get a drug on the market and to promote sales rather than to answer scientific questions.

In fact, randomised controlled trials, which involve testing new treatments against a control treatment or a placebo in a design that assumes the new treatment does not differ from placebo, began life as an effort to curb therapeutic overenthusiasm and as a means to debunk quacks. The very notion therefore that companies should use randomised controlled trials to market their compounds almost seems a contradiction in terms. In psychiatry and most of medicine, however, such trials have now become the fuel of bandwagons. They have been portrayed as offering evidence that transcends cultures, race, ethnicity, class, sex and age, evidence which defines good quality or rational medicine. When the majority of trials are carried out by pharmaceutical companies, and designed to suit their marketing requirements and when the data that result from company trials can be concealed from general scrutiny, there is cause for concern regarding such developments.

In the psychiatric domain there is an additional problem in that treatments are assessed solely on the basis of rating scales – that is how a patient responds to a series of questions about how he or she feels. Such scales are necessarily subjective and do not have the apparent objectivity of, for example, the fall of mercury in a sphygmomanometer used to assess the effects of an antihypertensive. Where in other areas of medicine, physicians, patients and outside observers can see the treatment effect, even though there may be doubt as to whether this effect will translate into a genuine benefit, in the case of psychiatry patients and others depend completely on the subjective judgement of physicians and companies.

For psychotropic drugs, the evidence that treatments have an effect is ordinarily drawn from scales rated only by physicians, specific to diseases codified in DSM-III. Apparently positive effects on such scales may in fact be produced by drug treatments doing more harm than good, at both an individual and population level. For example antidepressant treatments typically show short-term benefits in a few items such as sleep or agitation, without showing any benefit in the quality-of-life scales commonly completed by patients in the course of trials. In fact, there has been a comprehensive failure to publish measurements from quality-of-life domains completed in trials of SSRIs. Even if both patient and physician ratings indicated short-term benefits, it is by no means clear that the effects of a drug-induced dependence would not outweigh any benefits from short-term treatment in the longer term, but current trials are only designed to look for benefits after short exposure.

In summary, in this marketplace that claims to be following the evidence, a substantial amount of evidence remains unpublished and the data that are published come largely from trials that are inadequate. Even when the outcome measures used to decide when a treatment is working or not are the immediate onset benefits as rated by physicians only, in close to 50% of the trials of antidepressants submitted to regulators since the mid-1980s, the new treatment has failed to differentiate from placebo. It is difficult to resist the conclusion that our current state of knowledge in psychiatry is hardly sufficient to understand the true nature of the conditions we are treating nor whether our treatments are truly beneficial (see also Chapter 10, this volume). This ignorance or uncertainty is rarely admitted, however, nor does it prevent the strident claims that all will be well if practitioners simply practise according to the evidence.

LEVERAGE

The marketing of panic disorder by Upjohn marks the first major marketing of a treatment within the psychiatric profession through the means of sponsored conferences and publications written up to outline key positive research findings and to highlight key commercial messages. The success of Upjohn with panic disorder also demonstrated that a certain amount of funding strategically placed can be guaranteed to create interest among the editors of both academic and mainstream periodicals and magazines as well as in the makers of television programmes who recognise that panic attacks have become newsworthy. This is what I mean by leverage since the interest thus created leads to a wider acceptance of articles on the new illness, and even though many of these articles on panic disorder recommended psychotherapy rather than drug treatment, the wider recognition of panic disorder can only lead to increased drug sales. Selling disease entities or disease diagnoses in this way changes the perceptions of both prescribers and patients and even those of us who might never seek treatment as the language we use to describe nervous problems changes.

How is leverage obtained? The example of Upjohn and panic disorder brings out a number of obvious ways, including the sponsoring of scientific symposia, educational or financial support to attend meetings, research support, the purchasing and dissemination of thousands of copies of favourable scientific articles. In addition, through advisory and consultancy panels companies set up networks of friendship and influence. The obvious influences on experts are now supplemented by the influence of direct-to-consumer advertising which in the United States runs at billions of dollars per year.

But there are a number of less obvious methods of exerting influence in the marketplace. These involve the establishment of patient groups, the ghostwriting of the scientific literature, the selective publication of datasets and a set of negative marketing techniques that involve the intimidation of critics.

By the mid to late 1980s, the establishment of patient groups had become part of the market development programme for any new drug. Companies have recognised that user groups are a potent way to market new and costly treatments, which patients fear they might be denied on grounds of cost. In this way groups that were formerly hostile to physical as opposed to talking therapies in psychiatry have been brought onside. Patient groups also offer a ready supply of speakers for a variety of academic and non-academic meetings as well as individuals who can be accessed by the media, to deliver messages about the benefits of the drug in question.

There have also been a number of other developments through the 1990s that can best be characterised as negative marketing. In March 2000, the Hastings Center Reports carried a series of articles on the merits of using Prozac in people who appeared unhappy rather than depressed, two of which were supportive of such a use, two of which were not supportive and a fifth article by me, outlining the thesis argued here, namely that the effects of SSRIs extend far beyond any effects they may have on individuals taking the drugs. Following publication of my article, Eli Lilly, the makers of Prozac, which had been a significant donor to the Hastings Center, discontinued their support of the Center.

In April 2000, a book critical of Prozac, *Prozac Backlash* (Glenmullen, 2000), was published. Many media outlets such as the *Boston Globe* and *Newsday* in New York received unsolicited reviews of this book by senior figures in US psychiatry from public-relations agencies working for Lilly in different parts of the country. Where once journalists might have been sceptical of material fed to them by corporations, it seems standard practice for health and science correspondents to accept such material at face value – after all it commonly comes with the endorsement of senior scientific figures (Rampton and Stauber 2001).

Finally, there has been an emergence of ghostwriting. Almost since the inception of the modern pharmaceutical industry, companies have written articles for physicians. With the growth in scale of medical conferences, the phenomenon that is now called ghostwriting began to take on a clearer shape. A managerial restructuring of pharmaceutical companies in the 1980s led to the outsourcing of a range of functions such as the management of clinical trials and writing up of trial results and other reports. This outsourcing gave rise to clinical research organisations (CROs) and medical writing agencies.

Concerns about the ghostwriting of articles by such medical writing agencies began to be voiced in the early 1990s. Initial efforts to quantify its extent in peer-reviewed journals gave estimates that up to 11% of published articles might be ghostwritten. More recent estimates suggest that up to 50% of the literature in therapeutics is at present ghostwritten (Healy, 2003a). Where before it was thought that ghostwriting affected review articles in supplements to relatively obscure journals, it is now clear that it is at least as likely to affect the reporting of clinical trials in the most prestigious journals in medicine.

By means of ghostwriting, companies can make academics into opinion leaders, who appear to have authored more articles in the most prestigious journals than others and who present trial data at national and international meetings. However, the raw data remain the property of the pharmaceutical company protected by confidentiality clauses, which prevent these academic authors from sharing it with colleagues in the way that has been traditional in science.

THE GLOBAL MARKET

A number of rules apply to the prescription-only market described here; these include product patents, intellectual property agreements, pricing mechanisms and informal rules such as those that enable companies to argue that the data from clinical trials are at one and the same time both proprietary and scientific. These rules combine to create a market in which companies focus on producing blockbuster drugs, often for the treatment of lifestyle-linked risk factors, such as raised lipid levels or mild hypertension, rather than core diseases, while at the same time a range of common infectious disorders in the Third World, or less common disorders such as epilepsy or multiple sclerosis in the West, are no longer seen as likely to yield a sufficient return on investment to warrant research or drug development.

Although not likely to deliver treatments for diseases of global concern, this new market is one in which pharmaceutical companies are aiming at global influence. The transit from anxiety to depression and back outlined above illustrates some of the ways and means that can be used. The clinical trial data, stemming from small samples of Western patients, often recruited by advertisements, is held to provide valid evidence that should transform current practices from Japan to South America.

The same process is leading to a medicalisation of childhood distress. There are at present exponentially increasing rates of prescription of psychotropic drugs to children. What underpins the pharmaceuticalisation of child psychiatry? First many clinicians, recognising the existence of childhood nervous states, have argued for a continuity between these states and adult depression. It is legitimate to hope that someday if childhood depressions were detected early and treated effectively that adult depressions and a range of secondary disabilities such as substance misuse and job and marriage breakdown might be avoided. But given that in adult psychiatry, the availability of treatments has led to an apparent increase in the frequency of the conditions, which treatments are supposed to eliminate, the use of current treatments in children would seem unlikely to lead to an elimination of childhood distress.

Both clinicians and parents appear at present to be influenced by a change in culture that encourages the recognition of deviance and implies that intervention may make a difference (see Chapter 16, this volume). There are probably two reasons why both clinicians and parents might expect interventions to make a difference. One hinges on the role of clinical trials in creating an illusion that drugs actually work, and a second centres on a perception that more is known about neuroscience than is actually the case (see Chapter 10, this volume).

The rise of neuroscience is important for more than one reason. The obvious reason is that developments in neuroscience will someday underpin therapeutic developments – although they have not done so to date. But there is another way in which neuroscience feeds through into clinical practice. Both patients and clinicians are constantly involved in defining and redefining a common language to underpin therapeutic exchanges. For a large part of the twentieth century, psychodynamic thinking heavily influenced the language surrounding nervous problems. Patients expected doctors to talk about sexuality and complexes. The demise of psychoanalysis demonstrates how shallow this language was. Nevertheless this language spread far beyond the clinic, deep into mainstream culture, where it can best be described as a psychobabble. Shallow though it might have been, it is absolutely essential that some language underpins therapeutic exchanges. One of the features of the selling of the SSRIs was they brought in their train a widespread but loose use of neuroscientific terms in what can best be called a biobabble. Patients now expect their doctors to talk about lowered serotonin levels, and lead articles in periodicals and newspapers now casually refer to lowered serotonin levels. This is a language almost devoid of scientific meaning, as the recent advertising copy for SSRI anxiolytics demonstrates with its reference to lowered serotonin levels being the cause of anxiety, but this language is critical to what is essentially a social experiment on children, as well as to the extension of Western psychopharmacology to foreign cultures.

The changing climate and increasing recourse to psychotropic drugs has led to increasing company efforts to run trials in children. But these trials will not make clinical practice in the arena of childhood nervous disorders. What in fact the licences that have stemmed from these 'private' trials do is to permit companies to vigorously create markets for their products for children and teenagers. Companies have been enabled to medicalise childhood distress, and as the rapidly changing culture surrounding the management of such problems indicates, companies have the power to change cultures and to do so in astonishingly short periods of time. It is the same power that is leading to a homogenisation of psychiatric culture in all parts of the world that can afford modern drug treatments.

SCIENCE OR NOT?

In one sense it is hardly surprising that a divide exists between the advocacy of treatments and the actual outcomes for patients, since there has almost always been a chasm between medical aspirations and medical attainment, even long before modern times. However there is a further reason for the current discrepancy in that the key studies that underpin treatment are company trials appearing in the pages of the most prestigious journals in medicine. Company-controlled pharmacotherapy has now captured the main ground with increasing evidence of failures of correspondence between the published and the underlying raw data. In the case of the SSRIs, for instance, all the major companies have coded suicidal acts that did not occur on placebo as suicidal acts occurring on placebo; this appears to have happened because of an excess of suicidal acts on SSRIs compared with those occurring on placebo (Healy, 2003b). These data have since been published in mainstream articles.

The discrepancies between the published scientific articles and the raw data may have arisen because company personnel undertook the primary act of authorship and senior figures in good faith put their names to the text. Alternatively, these senior figures may have become, through the networks outlined above, friends of the company who were happy to manage the data as companies wish. These networks become problematic precisely when access to the raw data is denied to anyone who would wish to interrogate it further. On the one hand, these clinical trial data are claimed as proprietary and confidential. But on the other hand, companies and senior academics actively promote the data as scientific and increasingly compounds are marketed under the banners of science and evidence-based medicine.

PROSPECTS

In addition to providing the firmament on which human actions take place, biology can intrude forcibly into human history. One way in which this has happened has been through the infections that have played a role in the fates of peoples from the dawn of history to the present day. A second is through genetics and sex. A third and increasingly important way is through drugs. The products of the mapping of the human genome and other biological developments will all ultimately come to the marketplace in the form of drugs of one sort or another.

Biology is a source of variation rather than standardisation. When neuroscience has developed further it will almost certainly be clear that individuals differ greatly in the configurations of their serotonin systems and that the notion that there is a 'right' level of serotonin, as the current marketing of the SSRIs suggests, is misguided. Biological variation is in general dimensional rather than categorical; allied to the dimensions of height and weight with which we are all familiar, there are almost certainly a set of dimensions involving introversion, extraversion and neuroticism among others. But 'dimensions' cannot be sold as readily in the current prescription-only market, which is better suited to the delivery of treatments for categorical disorders, such as antibiotics for infections, regardless of an individual's prior constitutional type, psychosocial situation, ethnic background, or the bedside manner of the physician. And so a psychiatry that focussed heavily on notions of constitutional vulnerabilities for over a hundred years has given way to a psychiatry that deals with seemingly discrete infections of the mind such as panic disorder, social phobia or generalised anxiety disorder, and treats these disorders without any consideration as to whether an individual's constitutional type, social situation or ethnic background might make any difference to treatment and with ever-diminishing concern for the interaction between physician and patient.

The development of neuroscience, which at present seems at least in part shaped by goals set by pharmaceutical companies and the regulatory apparatus, may yet underpin alternative psychopharmaceutical arrangements that could escape the medical domain. More precise neuroimaging and pharmacogenetic technologies could help generate better quality outcomes with pharmaceuticals than we have at present. If this happens, psychopharmaceuticals could evolve into lifestyle or cosmetic agents rather than agents restricted to the treatment of diseases.

Such developments in the science base would pose an extraordinary set of problems for the current psychopharmaceutical marketplace. Pharmacogenetics for example offers the prospect of segmenting the marketplace by providing drugs tailored to individual genomes. While outcomes may be better, there will be fewer suitable candidates for 'treatment' and the possibility of developing a blockbuster will be correspondingly minimised. There are furthermore formidable ethical problems in running clinical trials that depend on screening and rejecting patients on genetic grounds.

While scientific developments may therefore ultimately generate a very different perception of and management of nervousness than has been the case in the past twenty years, nevertheless the example of our recent transit from anxiety to depression and back demonstrates how powerfully modern marketing techniques can use scraps of scientific data to build cultures that may in many respects be inimical to our welfare. It is not clear that developments in the science base will ever overcome the bias introduced by marketing but ultimately if there is to be trust in the market, there will have to be some basic correspondence between market claims and the evidence from science.

Further reading

Further background and discussion of the issues raised here is to be found in the books by Healy (1998, 2002, 2003a) and Rampton and Stauber (2001).

16 Education in the age of Ritalin

PAUL COOPER

Attention deficit/hyperactivity disorder (AD/HD) is a medical diagnosis applied to children and adults who are experiencing characteristic behavioural and cognitive difficulties in important aspects of their lives, for example in familial and personal relationships at school or work. The diagnosis attributes these difficulties to problems of impulse control, hyperactivity and inattention. It is claimed that these problems are caused primarily by dysfunctions in the frontal lobes of the brain and that there are predisposing genes. Currently the diagnosis is claimed to relate to between 2% and 5% of all children of compulsory school age in England and Wales.

THE DIAGNOSIS OF AD/HD

In 1968 the American Psychiatric Association (APA) produced the first standardised set of criteria for what was then called hyperkinetic reaction of childhood. This gave way in 1980 to attention deficit disorder with hyperactivity (ADDH), which was revised in 1987 to attention deficit disorder (ADD). A subsequent revision (American Psychiatric Association, 1994) produced the current diagnostic criteria for attention deficit/hyperactivity disorder (AD/HD). These changes in nomenclature reflect changing conceptualisations of the nature of the condition, with a shift away from an emphasis on causation to a continuing emphasis on behavioural symptoms as the defining characteristics of the condition. According to the APA, children with AD/HD fall into one of three main subtypes: predominantly inattentive and distracted,

The New Brain Sciences: Perils and Prospects, ed. D. Rees and S. Rose.
Published by Cambridge University Press.

predominantly hyperactive–impulsive, and combining hyperactivity with inattention and distractibility.

This shift in diagnostic criteria is also reflected in the alternative diagnosis of hyperkinetic disorders (HD), which is a diagnostic category of the World Health Organisation's *International Classification of Diseases*. The criteria of the World Health Organisation (WHO) (1990) are almost identical with those of the APA in terms of content but are more restricted in scope, requiring a higher proportion of potential symptoms to be present before diagnosis can be made. This includes a requirement that impulsiveness is always present, whereas this is not the case for the APA criteria. Also, there is a requirement in the WHO criteria that symptoms are generally pervasive, whereas for the APA criteria pervasiveness across only two situations is required. The WHO places greater emphasis on the need for the diagnosing clinician to observe symptoms, whilst the US *Diagnostic and Statistical Manual* (DSM-IV) allows greater reliance on reports of symptoms. The consequence of these differences is that the diagnosis of AD/HD is more inclusive than that of HD, with the latter producing prevalence rates in the UK of between 1% and 2% of the childhood population (National Institute for Clinical Excellence, 2000).

AD/HD can be seen as a subcategory of what the UK government classifies as 'emotional, behavioural and social difficulties' (EBSD). There are many surface similarities between AD/HD and what we might refer to as socialised EBSD – the result of some trauma in the child's life, such as family discord or break-up, poor parenting, social deprivation, abuse and bereavement. Of course, the stresses involved in having a child with AD/HD may give rise to, or exacerbate existing interpersonal difficulties in the family, making the untangling of external and internal causes difficult.

A key feature of the AD/HD diagnosis is that it should only be applied when the symptoms present themselves both in school and in the home, and when they have serious consequences for the quality of life of the child and/or the other people involved. After all, everyone displays some AD/HD characteristics some of the time, but it is only very few people for whom these characteristics are harmful to everyday functioning, to the extent that they create serious long-term difficulties in relationships with others and in carrying out domestic and work- or school-related tasks.

Children with AD/HD are often portrayed as being of average to high ability, but they disturb their parents and teachers because their classroom achievement is erratic, and often below their apparent level of ability. The child with AD/HD will often be a source of exasperation to the generally effective teacher. On occasions the child may show high levels of performance, a ready wit and an imagination of a high order but with erratic performance. In one form of the disorder he or she may often appear disengaged, easily distracted and lacking in motivation. The child may appear to be lazy or wasting obvious ability in favour of being oppositional and disruptive. This is the pupil who never seems to be in his or her seat, who is constantly bothering classmates, and can be relied upon for little other than being generally off-task. All categories can be frustrating to teach because of their apparent unpredictability, their failure to conform to expectations and their tendency not to learn from their mistakes.

In addition to the primary problems of inattentiveness, impulsiveness and hyperactivity, one reading of the AD/HD literature suggests that as the child gets older, the years of exasperation and blame can also lead to serious disaffection. Thus the child with AD/HD, if not handled appropriately, may not only continue to have difficulties in concentration and impulse control, but also develop an entirely reasonable antipathy to school and a lack of motivation to even try to overcome his or her difficulties.

ASSESSMENT AND INTERVENTION

There is a consensus in the clinical literature that AD/HD is a multifactorial problem requiring multimodal assessment and intervention (Barkley, 1990; Detweiler *et al.*, 1999). Recommended assessment procedures involve clinical interviews with parents and children, consultation with school personnel, and psychometric testing. Although the diagnosis is usually made by a physician, it is recommended that psychologists and educationists, as well as physicians, should be members of the assessment team, and that a wider range of professionals be drawn on in particular circumstances (British Psychological Society, 2000). The aim is to assess whether or not the symptoms have been accurately observed, to explore the possible range of causes for the symptoms, and to rule out immediate and preceding environmental factors (such as ineffective parenting or poor teaching) as primary causes. The diagnosis can be made if:

- The symptoms are found to be consistent with the diagnostic criteria.
- The symptoms are found to be a cause of significant disruption to social, emotional and academic development.
- The symptoms are found to be traceable to the individual's earliest experience of life.
- Immediately preceding environmental causes are judged not to account fully for the symptoms.

Medication is commonly prescribed for children who are deemed to show clinically significant levels of hyperactivity. In the USA between 2% and 2.5% of all school-aged children are prescribed some form of medication for hyperactivity, with over 90% of these being prescribed the psychostimulant medication methylphenidate, better known as Ritalin (Greenhill, 1998). This has to be compared with the much lower figure of less than 1% of children in the UK receiving similar pharmacological treatment (National Institute for Clinical Excellence, 2000). Methylphenidate is seen as an extremely safe medication, being non-addictive, with only mild side effects for the majority of users (such as sleep disturbance and appetite suppression), which can be controlled and often avoided through careful adjustment of the dosage after attention is paid to routine and regular reports from users, their parents and teachers. However, the very fact of medicating children for attentional and behavioural problems raises important concerns. It is essential to acknowledge that the clinical literature stresses that medication alone should not be regarded as a sufficient treatment but should always be part of a multimodal intervention approach which employs educational, behavioural, psychosocial, cognitive and environmental interventions. It must be said that there is sometimes, and maybe often, a serious gap between this informed advice and clinical practice. Lack of resources is never an acceptable excuse for inappropriate treatment. Medication should not be the default mode.

This multimodal approach is based on the assumption that the condition results from predispositions of the child that are shaped and exacerbated by the social environment. From this perspective medication is seen as creating a 'window of opportunity' that allows social and other influences to be brought to bear effectively. The child's behaviour is often misunderstood by parents, peers, teachers and other adults as a problem of motivation and volition rather than a hypersensitivity to circumstances that most people tolerate. By developing an understanding of these behavioural problems in terms of the cognitive and other theories of the aetiology of AD/HD, adults and others should be able to make more informed judgements about how best to approach and facilitate the child's positive development. Regarded in this way, the concept of AD/HD as a disorder can help reinforce the widely held view that apparent behavioural problems are often experienced by their perpetrators as reasonable responses to what they experience as difficult circumstances.

This last point should be of particular interest to teachers and educational psychologists, as well as parents and professionals such as social workers, whose job it is to support parents and families. The challenge is to learn to listen to the child and see the world through his or her eyes. They also need to find points of connection and cooperation with these children in order to provide the circumstances which will help to rebuild their often shattered sense of self and develop the emotional strength, self-confidence, trust in others and self-belief necessary to succeed socially and academically. This approach is of course beneficial to almost all children but for those with AD/HD it is often essential rather than simply desirable. This draws our attention to the often noted but sad fact that the vast majority of children are very capable of 'putting up' with circumstances that do not support positive personal development and are actually highly negative in their social, personal and educational effects. Children with AD/HD differ in that they cannot 'put up' in this way, but rather are the first to 'crack' under the strain of unreasonable conditions that some of us, too readily, take for granted.

As with other forms of socialised emotional behavioural and social difficulties, children with AD/HD experience difficulty in doing things that others find relatively easy, especially conforming to the kinds of behavioural expectations that are common in schools. The circumstances giving rise to their behavioural problems make it more difficult for them to exercise self-control, when they are compared to most other children of the same age. Such children have low self-esteem. Therefore, they often do not believe that they are capable of meeting the challenges that schools face them with and become trapped in a repeating cycle of failure. There are skills associated with good behaviour, just as there are skills associated with school subjects. Pupils with AD/HD and socialised EBSD need to be taught what good behaviour is and how to do it.

THE EVIDENCE BASE FOR AD/HD

The evidence suggests that individuals with AD/HD respond to the world in ways that are different from the general population; that is, they have different ways of cognitively processing and responding to the external world. The apparent level of resistance of these patterns of response to external influence, in the form of normal behavioural correction by usually competent and successful teachers and parents, is taken to imply deeper structural underpinnings to these cognitive problems.

Although there is a variety of cognitive theories of AD/HD, most are based on the assumption that cognitive dysfunctions are underpinned by neurological problems (Tannock, 1998). The basis for this assumption can be traced through a long line of research dating from the early years of the twentieth century which repeatedly indicates close similarities between the symptoms of AD/HD and those produced by injuries particularly in the prefrontal cortex of the brain. Other studies suggest a link between neurological damage affecting this part of the brain as a result of toxin exposure and AD/HD-type symptoms. Whilst recent research has added support to the neurological aspects of AD/HD, such research is far from conclusive, and has been at times inconsistent in its findings.

Neuroimaging studies, although limited, suggest that individuals with AD/HD tend to have smaller structures in those regions of the brain, particularly the striatal regions, which control movement and behaviour (Tannock, 1998; Barkley, 1997). These findings, however, like those from electroencephalography, leave us with a great many unanswered questions. The main problem is that they do not establish a causal link between the brain abnormalities and AD/HD, but merely co-occurence. The neurological basis for AD/HD, therefore, remains an interesting and promising hypothesis that is as yet unconfirmed.

AD/HD has been found to be more common in the biological relatives of children with AD/HD than it is in the biological relatives of children who do not have AD/HD. The problem with these studies is that it is difficult to control for environmental factors which family members often share and which may influence the development of AD/HD-type behaviours. This problem is addressed through twin and adoption studies which have repeatedly shown a much greater incidence of AD/HD among identical twins than among non-identical twins. Similarly, studies which compare the incidence of AD/HD among children and parents who are biologically related with that of children and parents where the child is adopted, have tended to support a degree of genetic involvement (Rutter, 2001). These findings are given further weight by molecular genetic research which has identified genetic abnormalities in the dopamine neurotransmission system in some children diagnosed with AD/HD.

Cognitive research has increasingly focussed on impulsiveness as the central feature of AD/HD, and the possibility that the cause lies in a frontal lobe abnormality resulting in an inability of the child to inhibit impulsive responses. Barkley (1997) proposes a model which suggests that neurologically based problems of response inhibition lead directly to problems in four major *executive functions* of the brain which are essential to effective self-regulation. The first executive function is *working memory*, impairment of which makes it difficult for individuals to retain and manipulate information for purposes of appraisal and planning. The second function is that of *internalised speech*. It is suggested that self-control is exerted through a process of self-talk, during which possible consequences and implications of behaviours are weighed up and internally 'discussed'. The third executive function is that of *motivational appraisal*. This system enables us to make decisions by providing us with information about the emotional associations generated by an impulse to act and the extent to which the impulse is likely to produce outcomes we find desirable. The final executive function is that of *reconstitution* or *behavioural synthesis* necessary for planning new and appropriate behaviours on the basis of past experience.

BEYOND BIOMEDICAL RESEARCH

The clash between the biomedical paradigm and the dominant educational framework for understanding learning and behavioural difficulties, which favours social–ecological models, does have practical consequences of some importance. In its extreme form, it gives rise to an outright rejection of the AD/HD diagnosis on the grounds that it is an invalid construct. Slee (1995: 74) expresses this anti-AD/HD view succinctly:

> The monism of locating the nature of [classroom] disruption in the neurological infrastructure of the child is myopic and convenient. As complex sites of interaction on a range of levels, classrooms provide opportunity for dysfunction across a number

> of fronts . . . The search for institutional dysfunction is ignored by the diagnosticians' probes. ADDS [*sic*] simply refines and extends the individualising and depoliticising of disruption in schools.

Slee's argument is a subset of the wider argument against medical labels for educational, social and psychological problems. In the words of Switzky *et al.* (1996: 2):

> Classes in the areas of human services (e.g. psychiatry and education) . . . are usually problematic. Evidence for this can be found in the fact that such socially constructed classes as schizophrenia, depression, learning disabilities, behavior disorders, along with mental retardation, are continually the subject of frantic (usually emotional and philosophical) efforts by committees of experts attempting to devise revised definitions that will win greater acceptance among practitioners and researchers than existing definitions.

The historical analysis on which Switzky's view is based argues that the application of pathological labels to individuals, despite the liberal intentions espoused by those applying the labels, has been associated with the exclusion of the labelled individuals from mainstream educational services, and their subjection to narrow and low-aspiring educational regimes.

The use of medication in the treatment of AD/HD has been a major focus for controversy. It is argued that the medicalisation of disruptive behaviour removes a sense of responsibility for the control of behaviour from the individual child, parents and teachers. This process is further exacerbated by the use of a medication that artificially suppresses impulsive behaviour, by physically removing the need to exert self-control. The biopsychological research agenda tends to assume that a major reason why individuals with AD/HD behave as they do is located within the individual who bears the AD/HD diagnosis (Rose, 1997). The fear is that such assumptions inevitably lead to a misdiagnosis of the nature of problems in many instances. In turn, such misdiagnosis disadvantages individuals whose behaviour is the product of unsympathetic or harmful environmental conditions.

Others, writing from the perspective of the educational practitioner, argue that medication undermines the rights and dignity of the children for whom it is prescribed, by disabling their sense of agency and depriving them of the right to choose how to behave. Baldwin (2000) echoes many of these arguments, referring to his own empirical research with parents of children with AD/HD, and drawing particular attention to findings which suggest that some children suffer adverse side effects from the medication, including addiction. Baldwin argues that the use of medication is more often a reflection of the financial vested interest of pharmaceutical manufacturers, rather than being based on clinical judgement. He further suggests that these vested interests have an unhealthy influence on research which appears to support the use of medication for AD/HD.

These criticisms cannot be taken lightly. From the educators' perspective they represent potential challenges to teachers' professionalism. If these negative perspectives on AD/HD are valid, then teachers who participate in the assessment process and engage in interdisciplinary cooperation with medical professionals in the treatment of AD/HD are colluding in a process of subjugation of children's fundamental human rights.

The unhelpful polarisation between AD/HD seen as a set of problems induced by biological factors and as problems generated by the environment is a crude nature-versus-nurture argument that contributes virtually nothing either to our understanding of AD/HD or of emotional behavioural and social problems in general. It does, however, tell us a lot about the tribalism of competing disciplines and professions.

Frith (1992) offers a model which helps us to understand the ways in which biopsychosocial factors may interact in developmental disorders. This describes biological causes leading to cognitive deficits and thence to behavioural manifestations (e.g. the behavioural symptoms of autism or AD/HD). The extent and indeed to some extent the nature of the behavioural manifestations are influenced by a set of social and psychological factors, namely: experience, maturation, compensation and motivation. Thus the extent to which the neurological problems result in behavioural and social dysfunction will be influenced by the individual's learning and experience, which may, for example, give the individual skills which enable him or her to compensate for cognitive deficits, or provide the individual with a high or low degree of motivation which in turn will affect his or her ability to cope. Clearly, the severity of the initial biological problem will vary, as will the nature of the individual's experience and environment. Thus in some cases biology will be more dominant than environment in the aetiology of the disorder, whilst in others environment will be more dominant than biology. Given that biology is heavily implicated in most prominent theories of the nature of AD/HD, whilst its precise function is still being debated, it seems only sensible to take the biopsychosocial perspective and recognise that in a given case it will always be very difficult to tease out the biological and psychosocial strands.

Increasing numbers of pupils in our schools are receiving medication for AD/HD (National Institute for Clinical Excellence, 2000). The explanation for this state of affairs is partly cultural. There has been a steady increase in the proportion of people exhibiting mental health problems and seeking mental health services since the middle of the last century, with the largest rate of increase being among the 12–24 age group. This, in turn, may be related to a breakdown in traditional support structures in families and communities. There is certainly evidence to suggest that far less stigma is associated with these issues than in earlier generations. In a recent international (thirteen country) study of the life concerns of adolescents and their associated coping strategies (Gibson-Kline, 1996), concerns about school grades, fear of failure at school and problems with behaviour in class were top of the list of concerns, except among the most economically disadvantaged, and for these school failure was the next concern on their list after their primary concerns relating to poverty. After education, family relationship problems were the next most commonly cited concerns. When it comes to coping strategies, self-reliant strategies are the most commonly cited, included 'trying harder' and 'planning'. 'Seeking help from others' comes lower down the list, and then the preferred 'others' are usually peers.

These findings contribute to an image of young people, throughout the world, as being increasingly stressed and increasingly isolated within their own peer culture. The school is a key site where AD/HD symptoms are often exacerbated. Schools are also places where, to some considerable degree, external political factors constrain teachers in the extent to which they can insulate pupils against the negative effects of curricula and performance demands. Children in Britain are among the most frequently tested pupils in the world, and pupils who do not do well in national tests are increasingly seen as a threat to development and survival of schools and the careers of individual teachers. It is not surprising, therefore, that there is evidence to suggest that some pupils take a highly pragmatic view of medication, accepting that it helps them to cope with educational circumstances that the informed observer might think they should have to cope with. This is one of the findings from a rare study of pupils' perceptions of the experience of AD/HD, which involved intensive, informant style interviews with sixteen adolescents diagnosed with AD/HD (Cooper and Shea, 1999).

Our study found that all pupils accepted the AD/HD diagnosis, and many of the pupils interviewed claimed that Ritalin made a major contribution to their academic performance through its influence on their abilities to concentrate, and the concomitant influence on work rate and ability to follow rules. Typical responses included:

> When I'm on it [Ritalin] I work harder, and I'm nicer, but when I'm out of school [and not on Ritalin] I'm sometimes silly, or I act stupid, or do things that I wouldn't really do if I was on Ritalin.
>
> [When I'm on Ritalin] I have more control over what I say.
>
> *(12-year-old girl)*

> When I'm taking Ritalin I'm calmer. I can study more and everything. And when I'm not I really can't concentrate or anything.
>
> *(Girl, 13)*

> I can concentrate better on Ritalin, I think like. I get on with my work more, and I don't talk so much.
>
> *(Boy, 14)*

> It makes me – Ritalin and Pemoline and things – they make me think first. I can think for myself anyway, but they make me think even better for myself.
>
> *(Boy, 15)*

There was one respondent in this study – a fifteen-year-old boy – who refused to take Ritalin and other drugs because 'they muck up your head'. His vehement response indicated that he saw Ritalin as a form of control that masked the incompetence of teaching staff. Whilst no other respondent was as negative as this boy about medication, there were many references to the downside of taking Ritalin. Some felt that the effects of the medication were primarily for the benefit of others, particularly parents. Mostly, however, respondents indicated a pragmatic attitude, seeing Ritalin as providing them with opportunities to engage in effective learning in circumstances that would not be conducive to their learning were they not receiving medication. The downside of the medication, cited by some respondents, is a loss of spontaneity:

> Sometimes I like it [Ritalin], but sometimes I don't . . . If I do take it when we didn't have school, I wouldn't want to go outside and play with my friends, or, I would just want to stay in home by myself and read a book or watch television or something.
>
> *(Girl, 15)*

Mostly these respondents' attitudes to Ritalin was to see it as a tool which they used for the purpose of enhancing their educational performance.

These findings accord with the view that disability labels and the interventions that sometimes accompany them are, in some cases, seen by those so labelled as helpful. This challenges conventional views about the stigmatising nature of such labels. In these circumstances, educators, and other professionals, sometimes find themselves working in and with circumstances that they would not choose, but which are preferred by their clients.

CONCLUSION

The rise of the AD/HD diagnosis has generated much controversy. Teachers, and others, will have different views about the desirability of the 'epidemic' of such medical diagnoses. The pragmatic response of many teachers has been to work with the AD/HD diagnosis, rather than against it. There are ways in which educators sometimes use the insights that come with the AD/HD concept to create educational environments that are responsive to the needs of all pupils in schools. This is not to say that there is not the potential for the diagnosis, and the medical treatment that often accompanies it, to be used in illegitimate and abusive ways. These concerns, however, have to be placed against the pragmatic view of some of the consumers of the AD/HD diagnosis and its medical treatment. For many of us it is highly regrettable when differences between people are construed as 'disorders', especially when it is clear that it is the environment that is disordered rather than the individual. On the other hand, we live in an increasingly individualist culture. The AD/HD phenomenon and the growing reliance on mental health services are among the inevitable consequences of this. It is frustrating to educationists to acknowledge the validity of Professor Basil Bernstein's dictum that 'education cannot compensate for society'.

Part V **Conclusion**

Conclusion

DAI REES AND BARBRO WESTERHOLM

ISSUES EMERGING

Steven Rose began this book with a commentary on the spectacular advances in the neurosciences over recent years, the claims for their implications, and possible aspirations for the future. The chapters that followed have explained and explored many of these aspects in some detail. We now approach this concluding chapter with somewhat different perspectives because, unlike other authors, we have no personal distinction or first-hand knowledge in modern neuroscience or any of the cognate disciplines brought into the discussion such as philosophy, law or social sciences. We come from other areas of science, and from the oversight of medical research linked to issues of public interest and government policy. One of us (DR) has been more concerned with practical applications and industry, and the other (BW) with the parliamentary process and with public sensitivities to ethics and welfare. It is natural for us to look, as it were, down the other end of the telescope to ask questions not so much about neuroscience, philosophy, law, or social sciences in themselves, but about how the new developments might impact on everyday life.

The guided tour of the new brain sciences has certainly made us think again about humanity's understanding of itself and what this means for the norms and dynamics of social behaviour, for example how criminal responsibility should be decided in law; about the scope we already have through the use of drugs and which it seems will soon be greatly expanded by genetic engineering and stem cell technology, to redesign not only ourselves and our children but other

The New Brain Sciences: Perils and Prospects, ed. D. Rees and S. Rose.
Published by Cambridge University Press.

people such as those that the politically powerful might think in need of 'improvement'; and about the moral questions as to whether, if, how, and when, these new abilities may be used. Although these large questions are not neatly separated, running into each other as they do, they have provided a useful sequence around which to structure the book. Our plan for this concluding chapter is to dip into the Parts in search of some answers, though not always taking the chapters in quite the order that the science dictated they should be arranged. We comment in our own language rather than the specialists' and, for better or worse, on the basis of personal viewpoints and values. We cannot do justice to all the important points made in all the individual contributions and aim only to pick out a number of take home messages that seem important to us for lay readers, and offer them with apologies to the expert contributors for the simplified and selective use of their material.

Freedom to change: stretching our comprehension

The freedom to choose and to commit to a sense of personal purpose has always been supposed to be at the root of life's meaning, as expressed, for example in the creation myths to explain human existence in early times. But science then burgeoned to evaporate so many mysteries and make it seem natural to some that mechanistic explanations should now be developed for all things including the workings of our own minds, an ambition reinforced by others in recent times by the achievements of modern neuroscience and genetics. Thus might we come to think of ourselves as glorified robots, even though in practice and at the same time we approach our own personal decisions as if we really do have choice – should we wage war? what should we have for lunch? Peter Lipton points out that it has always been easy to construct arguments in logic against the idea of free will, yet, watertight though these might seem, they make such nonsense of human experience that we are driven to accept that there must be limitations in a philosophical method which has somehow arrived at the denial of this quality that we value so much. Nor is the rejectionist case really strengthened by the successes of the new genetics and neurosciences since their witness merely brings forth more of the same type of evidence rather than any that is truly new and independent. Patrick Bateson points to excellent reasons why free will should have emerged in human evolution to endow the supreme advantage of deciding on actions after weighing the consequences, rather than following blind rote as other species do. Another negative position that seems to have permeated general thinking from the spirit of scientism abroad in our age is a penchant for reducing complex explanations to simple bottom lines, such as that a living creature is no more than the predictable product of its genes, that biology is no more than a working out of the laws of chemistry which in turn simply follow from the laws of physics. Mary Midgley's philosophical analysis shows how such reductionism simply does not work and that we can make sense of the world only in terms of complementary perspectives, each with an independent validity.

These insights can only heighten the sense of mystery at the workings of the human mind rather than encourage any idea that the remorseless march of the new brain sciences will soon arrive at any 'total explanation' or 'final synthesis' of them. On the contrary, the picture seems to expand and become ever more elaborate. Merlin Donald, for example, shows that the human mind is ever changing despite the relative constancy of its genetic inheritance, adding more and more capacities not only through practical inventions but also by developing new powers and modalities of communication and thinking. It is these that have provided the platform from which to subdue so much of the physical world and an increasing part of the biological one too, to create the glories of art, culture and philosophy, and to pursue the ambitions and megalomanias of political power and commercial adventure. These changes have been shaped and powered as much by what happens between people as within them as individuals, an idea that is developed for modern politics by Hilary Rose. She pushes the balance even further towards the importance of the social, showing the driving influence of the tensions between the splintered perceptions and interests of different sections of society – divided for example by class, gender and race – within which cooperation and consciousness have been enhanced by the social pressures. Even as individuals we do not appraise personal possibilities by reasoned calculation alone since our perceptions are framed by emotions, without which we would have no link between thought and the motivation to act.

Many of these issues and questions are brought to a head in the moral problems posed by possible applications of the neurosciences themselves, as Regine Kollek points out in her search for a framework for resolving such matters. Any code must be derived from human value judgements as well as informed by proper understanding of the science involved and (as we scientists would hope) by what science has to say about the nature of life. This building of bridges between independent perspectives is precisely the problem addressed by Mary Midgley, and is particularly difficult here because we hardly yet have the language to discuss the task. Regine Kollek argues for liberation from the familiar constraints of the scientific method in which we seek to pin down the relationships between observation and hypothesis in terms of one-to-one correlations, in favour of more flexible and imaginative forms of representation such as extended use of metaphor.

Neuroscience and the law: responsibility and pragmatism

Although we have argued for the reality of free will, people can find themselves in circumstances in which they are driven by forces within or outside themselves into actions for which they cannot be held responsible (see the chapters by Bateson, McCall Smith, Sedley and Lipton). It is a problem for the law to decide whether and when this can constitute a defence. Since, as we have argued, science has such good reason to be daunted by the challenge of understanding the human mind, it may come as no surprise to find our contributors very conservative about the prospects for useful criteria from science. As a High Court judge, Stephen Sedley argues that the law needs its own paradigms of human responsibility. His reasons are that most crimes are committed under pressure of circumstances, making responsibility a question of degree which requires a line to be drawn in a continuum between choice and compulsion; this line must serve the needs of social management which is no business of science (deterrence, protection of the public) as well as justice; and even if expert views could be useful in principle, they have been found in practice to be divided and subject to change if not fashion. Patrick Bateson adds that arguments from science may in any case be far from definitive since they usually do no more than point to one influence amongst many in a complex system of interactions, and mentions a famous example in which a court was led by scientific advice towards conclusions that now seem absurd. John Cornwell describes a court case which concerns possible side effects of the psychopharmaceutical drug Prozac, in which it seems that lawyers attempted to bamboozle with garbled ideas in neuroscience, only to find that the jury rejected their sophistry to fall back on commonsense ideas of human responsibility. Alexander McCall Smith explains the logic by which the law has developed its own framework, pointing out that that if the explanations of science had turned out to be clear-cut and deterministic, they would certainly have had to be heeded – but in fact no consensus of this type is emerging from the new brain sciences, other than for cases of severe mental derangement (recognised on other grounds anyway) and extremist ideas of the sort discussed by Lorraine Radford. Lorraine Radford adds, however, that criminality is a matter not only for the courts but for social policy as well, and we must ask whether the new understanding has any part to play here (of which more later).

Stewardship of the new brain sciences: look before leaping

We start with a caution for medicine similar to the caution for philosophy with which Part I began. It is risky to look for the quick translation of new knowledge into new medical treatments, just as it is unrealistic to hope that emerging insights into brain mechanisms will offer short cuts to new understanding of mind. Yadin Dudai uses the example of the cavalier introduction of temporal lobotomies for psychiatric disorders as an anti-paradigm for realising the potential which almost certainly exists in the longer run for genetic intervention, brain transplantation, or indeed any other irreversible modification of the brain. Similar forebodings come from Angus Clarke and David Healy for the genetic origins of human intelligence and other behaviour traits, and the development and marketing of drugs for nervous conditions and mental distress. They are concerned like Dudai about disasters from false confidence but have serious worries also (reinforced by precedents) about the scope for manipulating imperfect knowledge for political (e.g. racist) and commercial ends. Clarke and Radford are both troubled by the possibility that the new biology could be invoked to screen for individuals with potential to cause social problems, leading to policies which, again with imperfect knowledge, would find inappropriate targets – most likely in the underprivileged. For better or worse however, and as Paul Cooper describes, the diagnosis and treatment of attention deficit/hyperactivity disorder (AD/HD) as a medical condition seems to have found an accepted place in school education, with over 2% of school-aged children in the USA though rather less than 1% in the UK prescribed some form of medication which at least some of them profess to find helpful.

Returning to the more positive side of Dudai's theme, Helen Hodges, Iris Reuter and Helen Pilcher write from a biotech venture company founded to develop safe new treatments using genetically modified embryonic stem cells as brain transplants for neurodegenerative diseases which have little hope of other cure. They show how complex is the web of science and safety issues which have to be settled for their project to succeed. Since questions arise here about morality as well as safety, particularly in sourcing material from human embryos, Guido de Wert follows with an ethical analysis to expose logical inconsistencies in positions that oppose such research while on the other hand accepting the use of human embryos for infertility treatments; and also to question the arguments for shelving embryonic stem cells to explore the alternative potential of adult stem cells.

TOWARDS A WIDER DEBATE

Some of the issues on which we have touched have already surfaced in the media and elsewhere as matters of public interest – the hope, for example, for an effective therapy for Alzheimer's disease in our lifetime, the fear that the new knowledge will be used for designer babies, the question whether science's new understanding of humankind challenges our freedom in imagination, emotions and spirit. Such questions are not to be resolved in a few words, since each is the tip of an iceberg with technical, philosophical and moral dimensions beneath. Nor will answers emerge until other attitudes and policies have taken shape – on the scale and direction of research investment, the mood of opinion on means versus ends, the extent to which people submit to the tide of materialism. The answers – in other words – are not givens but will only follow from further debate – and the better the quality and the more inclusive this is, the longer any eventual consensus is likely to last. The process needs to be actively promoted and key principles of engagement to be observed; discussion must be genuinely two-way (lecturing, hectoring and blame games are probably unavoidable but unhelpful from any quarter); the process must involve scientists and the public both at large and within the social professional and political institutions, and recognise as Steven Rose has pointed out in his award lecture on science communication (2003), that this is no straightforward dialogue between crystalline positions. It is more like a cocktail party of exchanges between many different individuals drawn from two broad camps labelled 'science' and 'the public' with as much difference within as between them. An astronomer (even if he represents science policy in high places) might agree with an accountant and disagree with a biologist over stem cell research. A pensioner with a friend or partner debilitated by Alzheimer's disease might favour funding research, unlike the voter who would rather have tax cuts or the parent for whom priorities should be child care and school education, not to mention those others who stand aside from debate. Since this is all so chaotic and has so many interests and dimensions intermingled, the process is hardly amenable to systematic management. Willy-nilly, however, the processes of democratic decision-making will come to conclusions somehow – good, bad or indifferent. We who are concerned about the development and application of the new brain sciences can only, as best we can, provoke discussion and disseminate information which we trust is reliable. There are encouraging signs that the process is gaining momentum. Models of good practice are evolving in which interests and issues are brought together in informal and freewheeling ways, including 'laboratory open days', conferences, workshops, citizens' juries, displays, and science festivals at which advances and issues are put on show for wider discussion. An increasing number of such initiatives are taken by universities, research funding agencies and government departments, with more information networks set up for journalists to gain easier and quicker access to reliable and expert information on current developments, hopefully the better to inform the public about scientific findings and reduce the number of unfounded scare stories and heartless messages of false hope for disease.

The debate always becomes heated over ethical controversies. These may be difficult to identify early in the research process, and still harder to resolve when in that nebulous state outside the specific contexts that will make them concrete, but the sooner they are acknowledged the better. First anxieties might even be apparent at the choice of subject for scientific study and then develop right up to the application of the results. The general public has a strong interest that research is conducted according to value systems in society, divided though its opinions might be, for example by religion. Science usually moves faster than society finds comfortable, and is certainly doing so in biotechnology and the new brain sciences. Sometimes this gap closes as familiarity with the new ideas and possibilities makes them less frightening, as we have seen for the acceptability of organ transplantation and the notion of brain death, and might yet see for the therapeutic use of embryonic stem cells. But moral prohibitions may remain and indeed be final on other matters such as eugenics and the conditions under which human subjects may be recruited for medical research.

It is now widely if not universally accepted that the scientist who has the freedom and responsibility of defining the research project or problem for investigation should also be accountable to society through formal mechanisms for the way the work is conducted, especially when human or animal subjects are involved or there is the possibility of impacts on the environment. Increasingly too, formal oversight mechanisms are in place for standards of honesty and integrity in the conduct and reporting of research. The standards to be met may be embodied in a code of ethics such as that which has been in place at Uppsala University, Sweden, since 1981:

> The development of scientific knowledge – both fundamental research and applied research – influences the everyday life of people to an ever-increasing extent. This means that researchers working at universities and at research institutions outside of the universities, both in the public sphere and in the private sphere, must reflect over and actively work with ethical questions to a greater degree.

The conduct and ethics of research are increasingly included as an obligatory part of the education and training of scientists. Experiments with humans and human tissues are reviewed by independent committees or other arrangements, usually operating at the several levels of the local institution (e.g. university), research funding agency, and national government. All this may work well enough when the ethical problems are well recognised and public opinion already alerted, but there is still plenty of scope for surprises. Processes must therefore be in place to identify and evaluate new issues which arise as science advances. Examples already cited in this book include the work of the Nuffield Council on Bioethics (2002) mentioned by Clarke, Lipton and McCall Smith, and ad hoc enquiries by Government such as the expert study on policy for stem cell research commissioned by the UK Department of Health (2000) discussed by de Wert. The Swedish

Parliament during the 1980s and 1990s sponsored many investigations including possible legislation on gene technology, the adoption of the definition of brain death prompted by the need for organ transplants, and the scope of prenatal diagnosis for termination of pregnancy. The European Science Foundation (2003), a leading sponsor of the proceedings reported in this book, has conducted a number of ethical and policy studies such as the use of experimental animals for research, human stem cell research, genetically modified plants, and general practice and conduct in research and scholarship.

Laws and other legal rules codify the morality generally accepted in society, but it is inevitable that ethics should always be ahead of the law and ethical standpoints do change in the light of new knowledge. In any case, it is best to think twice before turning ethics into law. Sometimes legislation on research ethics, though introduced with best of intentions, has led to situations that governments have had cause to regret. For example, negative actions on DNA research and genetics in Sweden, Germany and the Netherlands in the 1970s caused scientists and the research to move abroad to countries more tolerant towards them. It is important to respond to public concern but all too easy to act with ill-thought-out measures in haste, only to repent at leisure.

To summarise the summary and look forward to specific measures, we conclude that the dilemmas from the new brain sciences are of a sort already familiar from science in the modern era, but in a particularly sharp and urgent form. Great benefits are promised for medical relief of the worldwide problems in mental health mentioned by Steven Rose in his introduction, and fresh insights into the human condition as illustrated by Merlin Donald. However, there are risks of mistakes from carelessness and overconfidence, and others of a sort inevitable in any learning process of application. More sinister are possibilities for misuse by false prophets and opportunists, keen to promote particular views about human nature, political interests in society, and commercial opportunities for profit. Which directions are developed and whether these will be entirely in the public interest depends on who seizes the initiative. Science needs partnerships with the public and with public and private institutions to work for the best futures, which can only be established by initiatives from all sides. We convened this seminar because it seems important that such topics are widely and proactively discussed – in advance rather than in the wake of emerging technologies. Given the degree of commonality in social and ethical mores in Europe and the interests of its institutions in the orderly development of policy, we felt that an approach at the level of the community of European scientists would be valuable. The agenda would need to serve a double purpose: on the one hand to think out what the new insights of science might imply for our understanding of what it means to be human and, on the other, to explore the questions of ethics, law and social policy triggered by the new technologies. Both challenges involve the context as well as the content of science, in other words perspectives looking in on science need to be considered along with the development of science as it looks out. Crucial to the first would be partnership with the academic disciplines of the social sciences and humanities, especially philosophy. The second would need a dynamic with policy debates in social and political institutions. Amongst the emerging European institutions, the European Science Foundation, which has already formed such linkages and indeed used them for other problems (European Science Foundation, 2003), would be well placed to guide the scientific community in these responsibilities, and our seminar concluded by proposing that the Foundation give high priority to developing a leadership strategy to address these needs, to help ensure that all the many constituencies with an interest in the development of neuroscience and neurotechnology can be proactively engaged in helping to prepare for the wider issues raised by present and future research. It has already communicated the concerns from these proceedings to the European Parliament with a recommendation to establish a permanent forum to engage the scientific community 'in keeping ethical issues under review surrounding new therapies which fall under the heading of brain interventions, whether by drugs, surgery, transplantation or gene manipulation'.

References

Abbott, A. (2001). Into the mind of a killer. *Nature*, **410**, 296–8.

American Psychiatric Association (1968). *Diagnostic and Statistical Manual of Mental Disorders*. Washington, DC: APA.

(1980). *Diagnostic and Statistical Manual of Mental Disorders*, 3rd edn. Washington, DC: APA.

(1994). *Diagnostic and Statistical Manual of Mental Disorders*, 4th edn. Washington, DC: APA.

Anderson, S. W., Bechara, A., Damasio, H., Tranel, D. and Damasio, A. R. (1999). Impairment of social and moral behaviour related to early damage in human prefrontal cortex. *Nature Neuroscience*, **2**, 1032–7.

Argyle, M. (1975). *Bodily Communication*. London: Methuen.

Ayer, A. J. (1954). Freedom and necessity. In G. Watson, ed., *Free Will*. Oxford: Oxford University Press, pp. 15–35.

Baddeley, A. (2000). The episodic buffer: a new component of working memory? *Trends in Cognitive Sciences*, **4**, 417–23.

Baldwin, S. (2000) (with Paul Cooper). Head to head: AD/HD. *The Psychologist*, **13**, 623–5.

Barkley, R. (1990). *AD/HD: A Handbook for Diagnosis and Treatment*. New York: Guilford.

(1997). *AD/HD and the Nature of Self Control*. New York: Guilford.

Bateson, P. and Martin, P. (2000). *Design for a Life: How Behaviour Develops*. London: Vintage.

Benjamin, J., Li, L., Patterson, C., *et al.* (1996). Population and familial association between the D4 dopamine receptor gene and measures of novelty seeking. *Nature Genetics*, **12**, 81–4.

Bickerton, D. (1990). *Language and Species*. Chicago: University of Chicago Press.

Bjorklund, A. and Lindvall, O. (2000). Cell replacement therapies for central nervous system disorders. *Nature Neuroscience*, **3**, 537–44.

Bjorklund, A., Sanchez-Pernaute, R., Chung, S., *et al.* (2002). Embryonic stem cells develop into functional dopaminergic neurons after transplantation in a Parkinson rat model. *Proceedings of the National Academy of Sciences of the USA*, **19**, 2344–9.

Blakemore, C. (1988). *The Mind Machine*. London: BBC Books.

Blank, R. H. (1999). *Brain Policy: How the New Neurosciences Will Change Our Lives and Our Politics*. Washington, DC: Georgetown University Press.

Boer, G. J. (2000). The network of European CNS transplantation and restoration (NECTAR): an introduction on the occasion of its tenth meeting. *Cell Transplantation*, **9**, 133–7.

Borlongan, C. V., Tajima, Y., Trojanowski, J. Q., Lee, V. M.-Y. and Sanberg, P. R. (1998). Transplantation of cryopreserved human embryonal carcinoma-derived neurons (NT2N cells) promote functional recovery in ischaemic rats. *Experimental Neurology*, **149**, 310–21.

British Psychological Society (2000). *AD/HD: Guidelines and Principles for Successful Multi-Agency Working*. Leicester: BPS.

Brunner. H. H., Nelen, M., Breakfield, X. O., Rogers, H. H. and van Oost, B. A. (1993). Abnormal behaviour associated with a point mutation in the structural gene for monoamine oxidase A. *Science*, **262**, 578–80.

Burt, T., Lisanby, S. H. and Sackeim, H. A. (2002). Neuropsychiatric applications of transcranial magnetic stimulation: a meta analysis. *International Journal of Neuropsychopharmacology*, **5**, 73–103.

Cane, P. (2002). *Responsibility in Law and Morality*. Oxford: Hart Publishing.

Caspi, A., McClay, J., Moffitt, T. E. *et al.* (2002). Role of genotype in the cycle of violence in maltreated children, *Science*, **297**, 851–4.

Chadwick, D. J. and Goode, J. A. (eds.) (2000). *Neural Transplantation in Neurodegenerative Disease: Current Status and New Directions*, Novartis Foundation Symposium no. 231. Chichester: John Wiley.

Chalmers, D. (1995). Facing up to the problem of consciousness. *Journal of Consciousness Studies*, **3**, 200–19.

Chizh, B. A., Headley, P. M. and Tzschentke, T. M. (2001). NMDA receptor antagonists as analgesics: focus on the NR2B subtype. *Trends in Pharmacological Sciences*, **22**, 636–42.

Clarke, A. (1995). Population screening for genetic susceptibility to disease. *British Medical Journal*, **311**, 35–8.

(1997a). The genetic dissection of multifactorial disease. In P. S. Harper and A. Clarke, eds., *Genetics, Society and Clinical Practice*. Oxford: Bios Scientific Publications, pp. 93–106.

(1997b). Limits to genetic research? Human diversity, intelligence and race. In P. S. Harper and A. Clarke, eds., *Genetics, Society and Clinical Practice*. Oxford: Bios Scientific Publications, pp. 207–18.

Cooper, P. and Shea, T. (1999). AD/HD from the inside: an empirical study of young people's perceptions of the experience of AD/HD. In P. Cooper and K. Bilton, eds., *AD/HD: Research, Practice and Opinion*. London: Whurr, pp. 223–45.

Cornwell, J. (1996). *The Power to Harm*. Harmondsworth: Penguin.

Crick, F. (1994). *The Astonishing Hypothesis*, New York: Simon and Schuster.

Criminal Statistics (2001). London: Home Office.

Dabbs, J., Carr, T., Frady, R. and Riad, J. (1995). Testosterone, crime and misbehaviour among 692 male prison inmates. *Personality and Individual Differences*, **18**, 62–70.

Dabbs, J., Riad, J. and Chance, S. (2001). Testosterone and ruthless homicide. *Personality and Individual Differences*, **31**, 599–603.

Daly, M. and Wilson, M. (1988). *Homicide*. New York: Aldine de Gruyter.

Damasio, A. R. (1994). *Descartes' Error*. New York: Grosset.

(1999). *The Feeling of What Happens: Body and Emotion in the Making of Consciousness*. New York: Harcourt Brace.

Dawkins, R. (1976). *The Selfish Gene*. Oxford: Oxford University Press.

de Wert, G. and Mummery, C. (2003). Human embryonic stem cells: research, ethics and policy. *Human Reproduction*, **18**, 672–82.

Deacon, T. (1996). *The Symbolic Species*. New York: Norton.

Dennett, D. C. (1991). *Consciousness Explained*. Boston, MA: Little, Brown.

Department of Health (2000). *Stem Cell Research: Medical Progress with Responsibility*. London: Department of Health. Available at http://www.doh.gov.uk/cegc/stemcellreport.htm

Detweiler, R., Hicks, M. and Hicks, A. (1999). A multimodal approach to the assessment of ADHD. In P. Cooper and K. Bilton, eds., *AD/HD: Research, Practice and Opinion*. London: Whurr, pp. 43–59.

Dipple, K. M. and McCabe, E. R. B. (2000). Phenotypes of patients with 'simple' Mendelian disorders are complex traits: thresholds, modifiers and systems dynamics (Invited Editorial). *American Journal of Human Genetics*, **66**, 1729–35.

Dobash, R., Dobash, R., Cavanagh, K. and Lewis, R. (2002). *Homicide in Britain: Risk Factors, Situational Contexts and Lethal Intentions*. Swindon: Economic and Social Research Council.

Draaisma, D. (2000). *Metaphors of Memory: A History of Ideas about the Mind*. Cambridge: Cambridge University Press.

Dudai, Y. (2002). *Memory from A to Z: Keywords, Concepts and Beyond*. Oxford: Oxford University Press.

Dunnett, S. B. and Bjorklund, A. (eds.) (2000). *Functional Neural Transplantation*, vol. 2, *Novel Cell Therapies for CNS Disorders*, Progress in Brain Research no. 127. Amsterdam: Elsevier.

Dutton, D. (2002). The neurobiology of abandonment homicide. *Aggression and Violent Behaviour*, **7**, 407–21.

Ebstein, R. P., Novick, O., Umansky, R., *et al.* (1996). Dopamine D4 (D4DR) exon III polymorphism associated with the human personality trait of novelty seeking. *Nature Genetics*, **12**, 78–80.

Eibl-Eibesfeldt, I. (1989). *Human Ethology*. New York: Aldine de Gruyter.

European Science Foundation (2003). *Science Policy Publications*. Available at http://www.esf.org/esf˙genericpage.php?language=0§ion=3&domain=0&genericpage=3

Fanon, F. (1967). *The Wretched of the Earth*. Harmondsworth: Penguin.

Fausto-Sterling, A. (2001). Beyond difference: feminism and evolutionary psychology. In H. Rose and S. Rose, eds., *Alas Poor Darwin: Arguments against Evolutionary Psychology*. London: Vintage, pp. 174–89.

Feinberg, J. (1984–8). *The Moral Limits of the Criminal Law*, 4 vols. Oxford: Oxford University Press.

Feuerstein, G. and Kollek, R. (1999). Flexibilisierung der Moral: Zum Verhältnis von biotechnischen Innovationen und ethischen Normen. In C. Honnegger, S. Hradil and F. Traxle, eds., *Grenzenlose Gesellschaft?* Opladen: Leske und Budrich, pp. 559–74.

Filley, C. M., Kelly, J. P. and Price, B. H. (2001). Violence and the brain: an urgent need for research. *Scientist*, **15**, 30–9.

Finnegan, J.-A. (1998). Study of behavioural phenotypes: goals and methodological considerations. *American Journal of Medical Genetics (Neuropsychiatric Genetics)*, **81**, 148–55.

Flanagan, O. and Rorty, A. O. (1993). *Identity, Character and Morality: Essays in Moral Psychology*. Cambridge, MA: MIT Press.

Fodor, J. (1998). The trouble with psychological Darwinism. *London Review of Books*, **20**, 2.

Frankfurt, H. (1971). Freedom of the will and the concept of a person. *Journal of Philosophy*, **lxviii**, 1, 5–20.

Freed, C. R., Greene, P. E., Breeze, R. E., *et al.* (2001). Transplantation of dopamine neurons for severe Parkinson's disease. *New England Journal of Medicine*, **344**, 701–9.

Freeman, W. J. (1999). *How Brains Make Up their Minds*. London: Weidenfeld and Nicolson.

Frith, U. (1992). Cognitive development and cognitive deficit. *The Psychologist*, **5**, 13–19.

Gelerenter, J., Kranzler, H., Coccaro, E., *et al.* (1997). D4 dopamine-receptor (DRD4) alleles and novelty seeking in substance-dependent, personality-disorder, and control subjects. *American Journal of Human Genetics*, **61**, 1144–52.

Gibson-Kline, J. (1996). *Adolescence: From Crisis to Coping*. Oxford: Butterworth.

Glenmullen, J. (2000). *Prozac Backlash*. New York: Simon and Schuster.

Goleman, D. (1996). *Emotional Intelligence*. London: Bloomsbury.

Gould, S. J. (1984). *The Mismeasure of Man*. Harmondsworth: Penguin.

Grafman, J. and Wassermann, E. (1999). Transcranial magnetic stimulation can measure and modulate learning and memory. *Neuropsychologia*, **37**, 159–67.

Gray, J. A., Hodges, H. and Sinden, J. D. (1999). Prospects for the clinical application of neural transplantation with the use of conditionally immortalized neuroepithelial stem cells. *Philosophical Transactions of the Royal Society, Series B*, **354**, 1407–21.

Gray, J. A., Grigoryan, G., Virley, D., *et al.* (2000). Conditionally immortalized multipotential and multifunctional neural stem cell lines as an approach to clinical transplantation. *Cell Transplantation*, **9**, 153–68.

Greenfield, S. (1997). *The Human Brain: A Guided Tour*. London: Weidenfeld and Nicolson.

Greenhill, L. (1998). Childhood ADHD: pharmacological treatments. In P. Nathan and M. Gorman, eds., *A Guide to Treatments that Work*. Oxford: Oxford University Press, pp. 42–64.

Guardian (2001). PC kills wife and sons with hammer. Tania Branigan, August 30.

(2002). Man jailed for 'forgotten' murder of wife. Rebecca Allison, July 24.

Harper, P. S. (1997). Genetic research and 'IQ'. In P. S. Harper and A. Clarke, eds., *Genetics, Society and Clinical Practice*. Oxford: Bios Scientific Publications, pp. 201–5.

Harper, P. S. and Clarke, A. (eds.) (1997). *Genetics, Society and Clinical Practice*. Oxford: Bios Scientific Publications.

Hart, H. L. A. (1963). *Law, Liberty, and Morality*. Oxford: Clarendon Press.

(1968). *Punishment and Responsibility: Essays in the Philosophy of Law*. Oxford: Clarendon Press.

Healy, D. (1991). The marketing of 5HT: anxiety or depression? *British Journal of Psychiatry*, **158**, 737–42.

(1998). *The Antidepressant Era*. Cambridge, MA: Harvard University Press.

(2002). *The Creation of Psychopharmacology*. Cambridge, MA: Harvard University Press.

(2003a). *Let Them Eat Prozac*. Toronto: Lorimer.

(2003b). Lines of evidence on the risks of suicide with selective serotonin reuptake inhibitors. *Psychotherapy and Psychosomatics*, **72**, 71–9.

Hodges, H., Sowinski, P., Virley, D., *et al.* (2000). Functional reconstruction of the hippocampus: fetal versus conditionally immortal neuroepithelial stem cell grafts. In D. J. Chadwick and J. A. Goode, eds., *Neural Transplantation in Neurodegenerative Disease: Current Status and New Directions*. Chichester: John Wiley, pp. 53–69.

Hodgson, D. (1991). *Mind Matters*. Oxford: Oxford University Press.

Horder, J. (1992). *Provocation and Responsibility*. Oxford: Oxford University Press.

Hume, D. (1748). *An Enquiry concerning Human Understanding*, ed. T. Beauchamp (1999). Oxford: Oxford University Press.

Humphrey, N. (1983). *Consciousness Regained: Chapters in the Development of Mind*. Oxford: Oxford University Press.

Jacobs, P., Bruton, M., Melville, M. M., Brittan, R. P. and McClermont, W. F. (1965). Aggressive behaviour, subnormality, and the XYY male. *Nature*, **208**, 1351–2.

Jeannerod, M. (1994). The representing brain: neural correlates of motor intention and imagery. *Behavioural and Brain Sciences*, **17**, 187–245.

Johnson, S. (1751). *The Need for General Knowledge*. In D. Greene, ed. (2000), *Samuel Johnson: The Major Works*. Oxford: Oxford University Press.

Karmiloff, K. and Karmiloff-Smith, A. (2001). *Pathways to Language: From Fetus to Adolescent*. Cambridge, MA: Harvard University Press.

Kelly, L. (1988). *Surviving Sexual Violence*. Oxford: Polity.

Kircher, T. T. J., Senior, C., Phillips, M. L. *et al.* (2001). Recognizing one's own face. *Cognition*, **78**, B1–B15.

Kosslyn, S. M. (1992). Cognitive neurosciences and the human self. In A. Harrington, ed., *So Human a Brain: Knowledge and Values in the Neurosciences*. Basel: Birkhäuser, pp. 37–56.

Kripke, S. (1980). *Naming and Necessity*. Cambridge, MA: Harvard University Press.

Kulwicki, A. (2002). The practice of honour crimes: a glimpse of DV in the Arab World. *Issues in Mental Health Nursing*, **23**, 77–85.

Lane, H. (1984). *When the Mind Hears*. New York: Random House.

LeDoux, J. E. (1996). *The Emotional Brain*. New York: Simon and Schuster.

Lees, S. (1997). *Ruling Passions: Sexual Violence. Reputation and the Law*. Buckingham: Open University Press.

Liberman, A. M., Cooper, F. S., Shankweiler, D. P. and Studder-Kennedy, M. (1967). Preception of the speech code. *Psychological Reviews*, **7**, 431–61.

Lindvall, O. (2000). Neural transplantation in Parkinson's disease. In D. J. Chadwick and J. A. Goode, eds., *Neural Transplantation in Neurodegenerative Disease: Current Status and New Directions*. Chichester: John Wiley, pp. 110–128.

Lindvall, O. and Hagell, P. (2000). Clinical observations after neural transplantation in Parkinson's disease. In S. B. Dunnett and A. Bjorklund, eds., *Functional Neural Transplantation*, vol. 2, *Novel Cell Therapies for CNS Disorders*. Amsterdam: Elsevier, pp. 299–320.

Lodge, D. (2002). *Consciousness and the Novel*. London: Secker and Warburg.

Lukács, G. S. (1971). *History and Class Consciousness*. London: Merlin.

Mackay, R. D. (1996). *Mental Condition Defences and the Criminal Law*. Oxford: Clarendon Press.

Marr, D. (1982). *Vision*. San Francisco: Freeman.

McAuley, F. (1993). *Insanity, Psychiatry and Criminal Responsibility*. Dublin: Round Hall.

Mednick, S., Gabrielli, W. and Hutchings, B. (2003). Genetic factors in the etiology of criminal behavior. In E. McLaughlin, J. Muncie and G. Hughes, eds., *Criminological Perspectives: Essential Readings*, 2nd edn. London: Open University/Sage, pp. 77–90.

Midgley, M. (1995). Reductive megalomania. In J. Cornwell, ed., *Nature's Imagination: The Frontiers of Scientific Vision*. Oxford: Oxford University Press, pp. 133–147.

(1996). One world, but a big one. *Journal of Consciousness Studies*, **3**, 500–14.

(2001). *Science and Poetry*. London: Routledge.

Moir, A. and Jessel, D. (1995). *A Mind to Crime*. London: Michael Joseph.

Mouzos, J. (1999). *Femicide: The Killing of Women in Australia 1989–1998*. Canberra: Australian Institute of Criminology.

National Institute for Clinical Excellence (2000). *Guidance on the Use of Methylphenidate for ADHD*. London: NICE.

Nuffield Council on Bioethics (2002). *Genetics and Human Behaviour: The Ethical Context*. London: Nuffield Council on Bioethics.

Paradis, K., Langford, G., Long, Z., *et al.* (1999). Search for cross-species transmission of porcine endogenous retrovirus in patients treated with living pig tissue. *Science*, **285**, 1236–41.

Pinker, S. (1994). *The Language Instinct*. New York: HarperCollins.

(1997). *How the Mind Works*. New York: Norton.

Plomin, R. (2001). The genetics of *g* in human and mouse. *Nature Reviews Neuroscience*, **2**, 136–41.

Plomin, R., DeFries, J. C., McClearn, G. E. and Rutter, M. (1997). *Behavioral Genetics*. New York: Freeman.

Plomin, R., Asbury, K. and Dunn, J. (2001). Why are children in the same family so different? Nonshared environment a decade later. *Canadian Journal of Psychiatry*, **46**, 225–33.

Premack, D. (1986). *Gavagai*. Cambridge, MA: MIT Press.

Pritchard, C. and Stroud, J. (2002). A reply to Helen Barnes' comment on child homicide. *British Journal of Social Work*, **32**, 369–73.

R. v. *Falconer* (1990) 171 CLR 30.

R. v. *Parks* (1992) 2 SCR 871.

R. v. *Quick* (1973) 1 QB 910, [1973] 3 All ER 347.

R. v. *Stone* (1999) 2 SCR 290.

Radford, J. and Russell, D. (1992). *Femicide: The Politics of Woman Killing*. Buckingham: Open University Press.

Rampton, S. and Stauber, J. (2001). *Trust Us, We're Experts!* New York: Tarcher-Putnam.

Reznek, L. (1998). *Evil or Ill: Justifying the Insanity Defence*. London: Routledge.

Richardson, A. (1969). *Mental Imagery*. New York: Springer.

Rose, H. (1994). *Love, Power and Knowledge: Towards a Feminist Transformation of the Sciences*. Cambridge: Polity.

Rose, H. and Rose, S. (eds.) (2000). *Alas Poor Darwin: Arguments against Evolutionary Psychology*. London: Jonathan Cape.

Rose, S. P. R. (1997). *Lifelines*. Harmondsworth: Penguin.

(2000). The future of the brain. *Biologist*, **47**, 96–9.

(2003). How to (or not to) communicate science. *Biochemical Society Transactions*, **31**, 307–12.

Rutter, M. (1996). Stress research: accomplishments and the tasks ahead. In R. Haggerty, L. Sherrod, N. Garmezy and M. Rutter, eds., *Stress, Risk and Resilience in Children and Adolescents*. Cambridge: Cambridge University Press, pp. 354–85.

(2001). Child psychiatry in the era following sequencing the genome. In F. Levy and D. Hay, eds., *Attention, Genes and ADHD*. Hove: Brunner-Routledge, pp. 225–48.

Schweinhart, L. J., Barnes, H. V. and Weikart, D. P. (1993). *Significant Benefits*. Ypsilanti, MI: High/Scope Press.

Scoville, W. B. (1954). The limbic lobe in man. *Journal of Neurosurgery*, **11**, 64–6.

Scoville, W. B. and Milner, B. (1957). Loss of recent memory after bilateral hippocampal lesions. *Journal of Neurology, Neurosurgery and Psychiatry*, **20**, 11–21.

Segal, L. (2003). Explaining male violence. In E. McLaughlin, J. Muncie and G. Hughes, eds., *Criminological Perspectives: Essential Readings*, 2nd edn. London: Open University/Sage, pp. 211–26.

Shapiro, C. and McCall Smith, A. (1997). *Forensic Aspects of Sleep*. Chichester: John Wiley.

Sheahan, D. (2000). Angles on panic. In D. Healy, ed., *The Psychopharmacologists*, vol. 3. London: Arnold, pp. 479–504.

Slee, R. (1995). *Changing Theories and Practices of Discipline*. London: Falmer.

Sokal, A. and Bricmont, J. (1999). *Intellectual Impostures*. London: Profile.

Stanley, L. (1992). *The Auto/Biographical I: The Theory and Practice of Feminist Auto/Biography*. Manchester: Manchester University Press.

Star, S. L. (1992). The skin, the skull and the self: towards a sociology of the brain. In A. Harrington, ed., *So Human a Brain: Knowledge and Values in the Neurosciences*. Basel: Birkhäuser, pp. 204–28.

Stocker, M. (1999). Responsibility and the abuse excuse. In E. F. Paul, F. Miller and J. Paul, eds., *Responsibility*. Cambridge: Cambridge University Press, pp. 175–200.

Sutherland, S. (1992). *Irrationality: The Enemy Within*. London: Constable.

Switzky, H., Greenspan, S. and Granfield, J. (1996). Adaptive behavior, everyday intelligence and the constitutive definition of mental retardation. *Advances in Special Education*, **10**, 1–24.

Tang, Y.-P., Shimizu, E., Dube, G. R., *et al.* (1999). Genetic enhancement of learning and memory in mice. *Science*, **401**, 63–9.

Tannock, R. (1998). AD/HD: advances in cognitive, neurobiological and genetic research. *Journal of Child Psychology and Psychiatry*, **39**, 65–99.

Thornhill, R. and Palmer, C. (2000a). Why men rape. *The Sciences*, Jan/Feb, 30–6.

(2000b). *A Natural History of Rape: Biological Bases of Sexual Coercion*. Cambridge, MA: MIT Press.

Tur, R. H. S. (1993). Subjectivism and objectivism: towards synthesis. In S. Shute, J. Gardner and J. Horder, eds., *Action and Value in the Criminal Law*. Oxford: Clarendon Press, pp. 213–37.

Turner, G. (1996). Intelligence and the X chromosome. *Lancet*, **347**, 1814–15.

van der Laan, L. J. W., Onions, D. E., Hering, B. J., *et al.* (2000). Infection by porcine endogenous retrovirus after islet xenotransplantation in SCID mice. *Nature*, **407**, 90–4.

van Inwagen, P. (1975). The incompatibility of free will and determinism. *Philosophical Studies*, **27**, 185–9. (Reprinted in Watson (1982), pp. 46–58.)

Vastag, B. (2001). Many say adult stem cell reports overplayed. *Journal of the American Medical Association*, **286**, 293.

Veizovic, T., Beech, J. S., Stroemer, R. P., Watson, W. P. and Hodges, H. (2001). Resolution of stroke deficits following contralateral grafts of conditionally immortal neuroepithelial stem cells. *Stroke*, **32**, 1012–19.

Vieraitis, L. and Williams, M. (2002). Assessing the impact of gender inequality on female homicide victimisation across US cities. *Violence against Women*, **8**, 35–63.

Walker, L. (1984). *The Battered Woman Syndrome*. New York: Springer.

Wasserman, D. (2001). Genetic predispositions to violent and antisocial behaviour: responsibility, character, and identity. In D. Wasserman and R. Wachbroit, eds., *Genetics and Criminal Behaviour*. Cambridge: Cambridge University Press, pp. 303–27.

Wei, F., Wang, G.-D., Kerchner, G. A., *et al.* (2001). Genetic enhancement of inflammatory pain by forebrain NR2B overexpression. *Nature Neuroscience*, **4**, 164–9.

Wilson, M. and Daly, M. (1999). Lethal and non-lethal violence against wives and the evolutionary psychology of male sexual proprietariness. In Dobash, R. and Dobash, R., eds., *Rethinking Violence against Women*. London: Sage, pp. 199–230.

World Health Organisation (1990). *International Classification of Diseases*, 10th edn. Geneva: WHO.

Yoshikawa, H. (1995). Long-term effects of early childhood programs on social outcomes and delinquency. *The Future of Children: Long-Term Outcomes of Early Childhood Programs*, **5**, 51–75.

Ziman, J. (1978). *Reliable Knowledge: An Exploration of the Grounds for Belief in Science*. Cambridge: Cambridge University Press.

Index